Giovanni Salvi

## Explicit solvent effects on protein physics

Giovanni Salvi

# Explicit solvent effects on protein physics

## Modeling lattice proteins in aqueous environments

Südwestdeutscher Verlag für Hochschulschriften

**Impressum/Imprint (nur für Deutschland/ only for Germany)**
Bibliografische Information der Deutschen Nationalbibliothek: Die Deutsche Nationalbibliothek verzeichnet diese Publikation in der Deutschen Nationalbibliografie; detaillierte bibliografische Daten sind im Internet über http://dnb.d-nb.de abrufbar.

Alle in diesem Buch genannten Marken und Produktnamen unterliegen warenzeichen-, marken- oder patentrechtlichem Schutz bzw. sind Warenzeichen oder eingetragene Warenzeichen der jeweiligen Inhaber. Die Wiedergabe von Marken, Produktnamen, Gebrauchsnamen, Handelsnamen, Warenbezeichnungen u.s.w. in diesem Werk berechtigt auch ohne besondere Kennzeichnung nicht zu der Annahme, dass solche Namen im Sinne der Warenzeichen- und Markenschutzgesetzgebung als frei zu betrachten wären und daher von jedermann benutzt werden dürften.

Verlag: Südwestdeutscher Verlag für Hochschulschriften Aktiengesellschaft & Co. KG
Dudweiler Landstr. 99, 66123 Saarbrücken, Deutschland
Telefon +49 681 37 20 271-1, Telefax +49 681 37 20 271-0, Email: info@svh-verlag.de
Zugl.: Lausane, EPFL, 2004

Herstellung in Deutschland:
Schaltungsdienst Lange o.H.G., Berlin
Books on Demand GmbH, Norderstedt
Reha GmbH, Saarbrücken
Amazon Distribution GmbH, Leipzig
ISBN: 978-3-8381-0771-4

**Imprint (only for USA, GB)**
Bibliographic information published by the Deutsche Nationalbibliothek: The Deutsche Nationalbibliothek lists this publication in the Deutsche Nationalbibliografie; detailed bibliographic data are available in the Internet at http://dnb.d-nb.de.

Any brand names and product names mentioned in this book are subject to trademark, brand or patent protection and are trademarks or registered trademarks of their respective holders. The use of brand names, product names, common names, trade names, product descriptions etc. even without a particular marking in this works is in no way to be construed to mean that such names may be regarded as unrestricted in respect of trademark and brand protection legislation and could thus be used by anyone.

Publisher:
Südwestdeutscher Verlag für Hochschulschriften Aktiengesellschaft & Co. KG
Dudweiler Landstr. 99, 66123 Saarbrücken, Germany
Phone +49 681 37 20 271-1, Fax +49 681 37 20 271-0, Email: info@svh-verlag.de

Copyright © 2009 by the author and Südwestdeutscher Verlag für Hochschulschriften Aktiengesellschaft & Co. KG and licensors
All rights reserved. Saarbrücken 2009

Printed in the U.S.A.
Printed in the U.K. by (see last page)
ISBN: 978-3-8381-0771-4

# Contents

**1 Introduction** — 1
   1.1 Proteins — 1
   1.2 Protein Structures — 2
   1.3 Protein Folding and Protein Design — 5
   1.4 Solvent effects and hydrophobicity — 10
   1.5 Cosolvents — 15
   1.6 Overview — 17

**2 The models** — 19
   2.1 The HP model — 21
   2.2 The HPW model — 24

**3 Protein Design** — 31
   3.1 *Good* sequences are needed — 31
   3.2 Compactness of the native states — 34
   3.3 Protein Folds and Designability — 36
   3.4 Cold and warm denaturation — 38
   3.5 Statistical properties of the sequences — 42
      3.5.1 Hydrophobic amino-acid concentration — 42
      3.5.2 The Run Test — 44
   3.6 Overall Stability Criterion — 47

**4 Chaotropic Cosolvent Effects** — 51
   4.1 The modified HPW model — 52
   4.2 Protein stability — 55
   4.3 The condensation of chaotropic molecules — 57
   4.4 Derivation of the $m$-value — 61
   4.5 The $(T, c)$ phase diagram — 65

**5 Solvent effects on effective interactions** — 69
   5.1 Effective interactions on real proteins — 69
   5.2 Effective interactions on lattice models — 72
      5.2.1 The perceptron — 73
      5.2.2 The perceptron on lattice proteins — 75
      5.2.3 The perceptron on the HP model — 78
      5.2.4 The perceptron on the HPW model — 80

**6 Summary** — 93

# Bibliography

# Chapter 1
# Introduction

Protein folding stands as one of the major interdisciplinary challenges of the last fifteen years, involving biology, chemistry, medicine and physics. In this book solvent effects, and the related *hydrophobic effect*, on proteins are investigated. Using a simple lattice model of proteins, in which the solvent is semi-explicitly taken into account, thermodynamical quantities can be investigated and the crucial role the solvent plays in protein folding can be demonstrated.

## 1.1 Proteins

Proteins are macromolecules that play a central role in most biological processes [1]. These molecules, which constitute a major fraction of the mass of all organisms, are responsible for catalyzing and regulating biochemical reactions, transporting molecules, as well as forming the basis of structures such as hair or skin (Fig. 1.1).

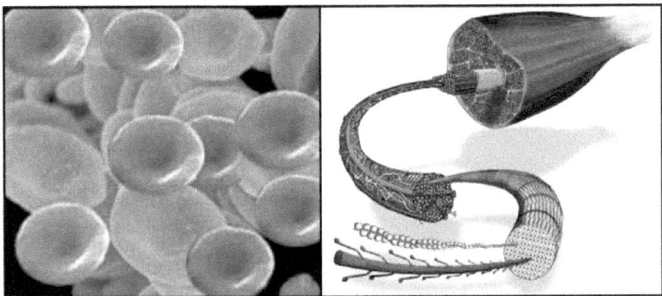

Figure 1.1: Two well-known protein functions. In the blood red cells, a protein called Hemoglobin is responsible for the oxygen transport through the whole body and gives the characteristic red color to the red cell [2]. In the muscles, two proteins, the Actin and the Myosin, are responsible for the muscle contractions and are therefore the cause of every muscle movement in the body [3].

A protein is a long linear polymer, composed by a specific sequence of monomers, that are the 20 naturally occurring *amino-acids* (also referred to as *residues*). The chemical structure of these monomeric units can be divided in two components: a common part, composed by an amino group and a carboxylic acid group bound together by a carbon atom, and another part, referred to as *side chain* (or *R group*), that characterizes each residue (Fig. 1.2). The variety of these side chains ranges from simple units such as a single hydrogen atom in the case of the Glycine amino-acid, reaching structures as complex as aromatic groups in the case of Tyrosine or Tryptophan residues.
Each amino-acid has specific chemical properties. In particular, one can class the residues into two main groups, distinguishing *polar* from *non-polar* amino-acids; such difference, as we shall see, plays a fundamental role for the protein folding process.

Residues have to be polymerized in order to form a protein. In the nucleus of a cell, this task is performed by the ribosome: a very complex molecule composed both by ribosomal RNA (rRNA) and by proteins. The information about the amino-acid sequence of any protein is coded in the DNA: once the cell needs a specific protein, its related part of DNA is "translated" into messenger RNA (mRNA), a single-stranded molecule. From this molecule, the ribosome is able to decode the information needed to synthesize the correct protein. Transfer RNA (tRNA) carries amino-acids to the ribosome, to enable it to stick the needed residue on the protein that is being formed following the code given by the tRNA. Through elimination of a water molecule, the ribosome condenses pairwise amino-acids: the iteration of this process gives rise to long linear chains (Fig. 1.2).

Eventually, the resulting chains of residues collapse in a compact and stable three-dimensional structure, forming a macromolecule (*i.e.* the protein) with specific sterical and chemical properties (Fig. 1.3).

Since a single protein is a chain of tens to thousands of residues, an astronomic number of different combinations is available to create, at least in principle, an amino-acid chain. Yet, through selective pressure, only a small part of them has been selected, favoring proteins with very specific properties. In particular, proteins with suitable amino-acids on their surfaces have been retained, since the chemical properties of a protein are mainly determined by the kind of residues on its surface: indeed it is able to react with other molecules, and therefore to perform its function thanks to its surface chemical features (the surface regions where interactions take place are referred to as *active sites*).

Given the importance of proteins for life and their enormous diversity, it is not surprising that protein research stands nowadays as one of the major interdisciplinary challenges, involving biology, chemistry, medicine and physics.

## 1.2 Protein Structures

The knowledge of the three-dimensional structure of a protein is a basic requirement in order to understand its properties and function. To be more precise, only quantitative information about shape, size and, in particular, position and orientation of the residues at the surface allows to study a given protein with the appropriate *in vitro* experiments.

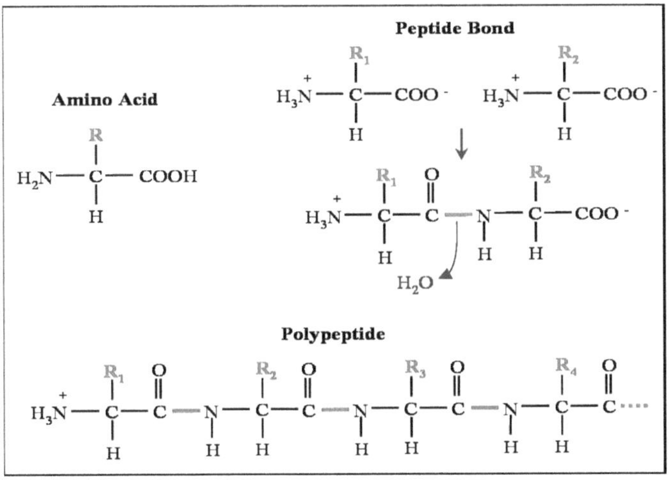

Figure 1.2: Schematic representation of protein polymerization. Two amino-acids are dimerized through the elimination of a water molecule, creating the so-called *peptide bond* (the CO-NH linkage). Further residues are added by the same procedure creating long chains which finally fold in a compact structure.

These analyses may then provide the necessary information describing the protein interactions with other molecules. Nowadays, two main techniques have been developed to achieve this task: *X-ray crystallography* and *nuclear magnetic resonance spectroscopy*.

Despite being the first, X-ray crystallography is still the most used reliable method to obtain protein structures. The idea underlying this method is the application to proteins of the well-known crystallographic techniques. In order to obtain a suitable crystal of a given protein, the latter has first to be purified and put in solution. Once the protein has attained a reasonable state of purity, the protein solution is brought past its saturation point and the protein may precipitate from the solution in crystalline form. Finally, if the size of the resulting crystal is large enough, it may be suitable for X-ray crystallographic analysis (Fig. 1.4).

However, despite the constant improvement of the techniques used in the crystallization process and the introduction of automatization procedures, obtaining a valuable protein crystal is still an arduous task. Indeed, even in the best case, weeks (or months) of efforts are needed to obtain a crystal, while in the worst case all attempts simply fail, since the protein may not crystallize or because the grown crystal may not reach a suitable size for a reliable X-ray experiment. In addition, a more crucial problem exists with the application of X-ray crystallography to the study of macromolecules, namely their *flexibility*: it has recently been shown, that in some cases there is a considerable difference between the three-dimensional structure obtained by crystallization methods and the real structure of the protein *in vivo*. For example, in the case of the Immunoglobulin G (IgG)

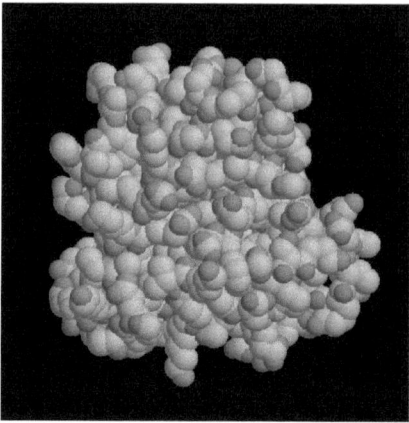

Figure 1.3: Atomic representation of the Myoglobine protein, the very first known protein structure. The color code is: gray for Carbon groups, blue for Nitrogen groups, red for Oxygen groups and pink for lateral side chains [4].

protein, it has been observed that two of its subunits are highly flexible and therefore may induce large-scale conformational differences in the protein structure. It follows that the IgG structure obtained by X-ray crystallography, rather as being the thermodynamical equilibrium conformation, is only one structure among a large number of different conformations, each of them having roughly the same probability of occurrence [6].

About ten years ago nuclear magnetic resonance (NMR) spectroscopy was employed for the first time to resolve the structure of a protein. This technique has the considerable advantage over X-ray crystallography in that experiments are performed in aqueous solution, *i.e.* the same conditions as *in vivo*, eliminating the problem of possible discrepancies between the observed structure and the real structure of proteins. The idea of this method is to measure the precession of proton spins (in presence of an external magnetic field), since their frequency depends on the type of atom the proton is bound to and on the local chemical environment. Therefore, it is possible, at least in principle, to determine the molecule structure by analyzing the frequency spectrum of precession. In practice however, it is established that protons belonging to different amino-acids have degenerate frequencies (becoming therefore indistinguishable). Moreover, the number of atoms involved increases with the size of the protein. As a consequence, this method can only be successfully applied to proteins of less than 30-40 kD. However, this technique is very efficient for small proteins and is also useful for the investigation of parts of proteins, since, as opposite to the crystallization method, there is no need to deal with the whole structure at once.

Nowadays, more than 20.000 protein structures are stored in the *Protein Data Bank* (PDB) [4], of which around 3.000 were resolved by NMR spectroscopy and the rest by

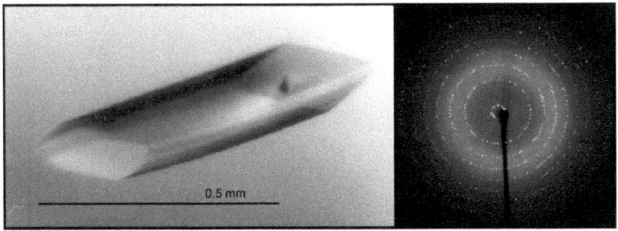

Figure 1.4: Example of a (RGS14) protein crystal and its related diffraction pattern. Once the crystal has reached a suitable size, it is possible to obtain a X-ray diffraction picture, which is then elaborated in order to extrapolate the structure of the protein [5].

X-ray crystallography. As a matter of fact, the size of the PDB grows exponentially since its creation in 1972, as more and more structures are added every day.

## 1.3 Protein Folding and Protein Design

A completely different approach to the acquisition of protein structures is followed in the *protein folding* problem. The aim of protein folding is the investigation of the folding process of the initial unfolded amino-acid polymer toward the final three-dimensional structure.

One of the most important contribution to the understanding of the folding process was given by C. B. Anfinsen *et al.* in the early seventies, while dealing with a specific protein: the *ribonuclease* molecule [7]. During their experiments, these proteins in solution were first denaturated (*i.e.* the original protein structure was disrupted) by adding urea in the solution. The urea was then eliminated and, after equilibration, the structure of the amino-acid chains was investigated. It turned out that the structure of the ribonuclease molecules after this process was indistinguishable from the original one. This important result led to the establishment of the originally called *thermodynamic hypothesis*: the *native state* (*i.e.* the *in vivo* structure) of a protein, at physiological conditions, is the one possessing the lowest free energy among all other possible conformations. This statement, which is nowadays accepted as the *Anfinsen's dogma*, implies that the three-dimensional structure of a protein is completely encoded in its own amino-acid sequence. As a consequence, the knowledge of the latter is sufficient, at least in principle, to determine the native state.

The equivalence between the sequence of residues along the chain and the native state of a protein allows to introduce two new definitions: the amino-acid sequence is also referred to as *primary structure* of a protein, while the three-dimensional structure is normally called *tertiary structure*. In addition, proteins, during their folding, create particular local structures, such as helices, known as $\alpha$-*helices* and planar arrangements of residues called $\beta$-*sheets*. These structures are referred to as the *secondary structures* of the protein [1]. These three different protein representations are pictorially shown in Fig. 1.5.

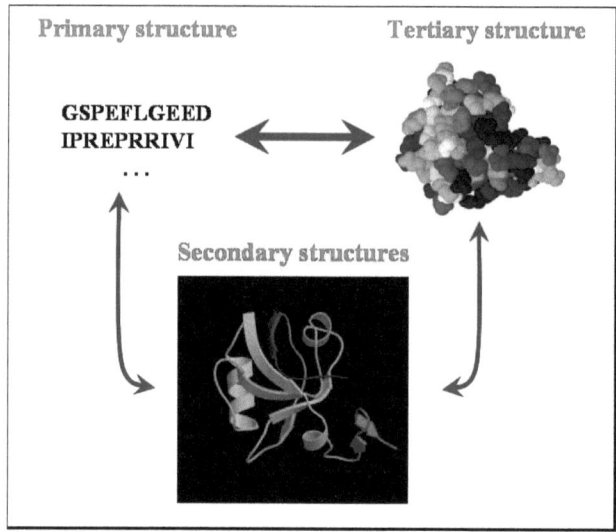

Figure 1.5: Schematic representation of the different levels of structural organization of proteins. A given amino-acid sequence (primary structure) always folds, at physiological conditions, into the same three-dimensional conformation (tertiary structure). In other words, the native state of a protein is completely encoded in its amino-acid sequence. As for the secondary structures, the picture shows an $\alpha$-helix in green and some $\beta$-sheets, represented by arrows of different colors.

A further fundamental advancement to the understanding of the protein folding problem has been contributed by the solution of the so-called *Levinthal's paradox* [8, 9]. It had been argued for a long time that, even if the native state is the global minimum of the free energy, it is not realistic to expect the folding of a protein to be only driven by a free energy gradient. This argument was based on the simple observation that, the huge number of atoms that makes up a protein imply an enormous number of degrees of freedom for the system. Consequently, it would be impossible for the chain of residues to explore the whole phase space in order to reach the global minimum of free energy. It has actually been calculated that even for a very short protein made of 40 amino-acids it would take about $10^{18}$ years to examine all possible conformations. This period of time is obviously much longer than the present age of the Universe. Therefore, if the folding process is only lead by the free energy, as claimed by the Anfinsen's dogma, how is it possible that the typical folding time of proteins does not exceed a few seconds? The answer to this paradox was given by K. A. Dill in 1999, while investigating the shape of the free energy landscape [10]. It turned out that the latter, instead of being a very rough landscape with a huge number of local minima, as expected from the complexity of the system, is much more similar to a *funnel* (Fig. 1.6).

It follows that the native state has to lie at the very bottom of the funnel, since this is the state with the lowest free energy. Therefore, the chain of residues folds following the walls of the funnel and, even if the folding process may be occasionally slowed down by kinetic traps (the smooth walls of the funnel in Fig. 1.6 are not realistic), the number of possible conformations decreases rapidly during the folding and the native state is reached in a reasonable time, thus disproving the Levinthal's paradox.

The two above important results led to the current main challenge in the understanding of the protein folding problem: *finding a suitable Hamiltonian in order to obtain the native state of a protein from the simple knowledge of its primary structure*. In fact, if a Hamiltonian describing the system could be found, the free energy of the system could be easily computed and the native state of the investigated protein would "simply" be the conformation with the lowest free energy.

In principle, protein folding could be studied using *all-atom Molecular Dynamics* (MD). However, even with the fastest present computer clusters it is not possible to simulate the entire folding process of a generic protein. Indeed, it is important to stress that for this kind of simulations, both the atoms of the protein and the atoms of the solvent have to be taken into account, so that the number of molecules involved is enormous. Yet, Molecular Dynamics is a powerful technique for the investigation of protein functions, following up to *nanoseconds* of the behavior of folded proteins (folding processes last $10^{-3} - 10^1$ seconds!). Moreover, MD has also recently been successful applied for the complete folding of small proteins [11].

Because of the impossibility to deal directly with the detailed description of the system, people have introduced models that reduce the complexity of the process to a manageable level. One usual strategy is to determine effective contact interactions between aminoacids for coarse-grained models of proteins [12, 13, 14]. The idea is to understand, by investigating known protein structures, which couples of residues are favorable or unfavorable. For example, two amino-acids with non-polar side chains, such as Leucine or Glycine, have a propensity to stay close to each other and, therefore, an attractive interaction is associated to this couple. On the contrary, couples made by charged polar residues, such as Lysine or Arginine, will be characterized by a repulsive interaction.

Proteins are made by 20 different kinds of amino-acids. Hence, all contact interactions can be encoded in a $20 \times 20$ matrix [12, 13, 14], the native state corresponding to the minimum of a semi-empirical Hamiltonian depending on such pair interaction energies [15]. In order to determine these interactions, several techniques have been developed, such as the *Perceptron* algorithm [16], where a simple neural network tries to learn contact interactions from a given training set of sequences and structures (as, for example, from the PDB database) [17].

However, the application of the effective interaction method has several limitations. It turns out that there is not a unique and general interaction matrix, but, on the contrary, different proteins are described by different interaction matrices [18]. Depending on which

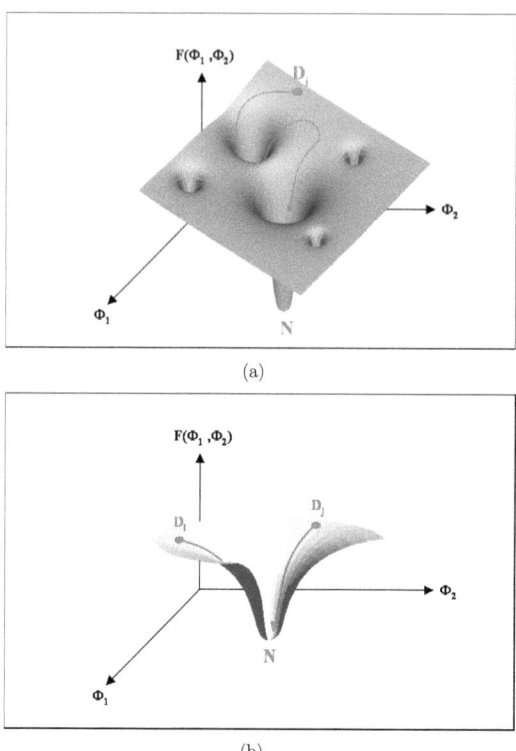

Figure 1.6: Schematic representation of the free energy landscape of proteins. (a) *The Levinthal's paradox*. It was believed that, because of the enormous number of degrees of freedom of the system, it would be impossible for a chain of residues to explore the whole phase space in order to reach the global minimum of free energy. (b) In reality, the free energy landscape of a protein possesses a funnel-like shape. Regardless of the initial denaturated conformation, the amino-acid chain folds downhill the walls of the funnel. Through this process the number of accessible conformations decreases drastically, forcing the protein to fold in its native state which lies at the bottom of the funnel.

set of proteins is chosen for the computation of the interaction matrix, different values are obtained for the same pairwise interactions. This discrepancy among computed interaction matrices implies that the interactions are efficient in finding the correct native state of the proteins used for the learning process (and, in some special cases, of few others), but they are not able to correctly fold a generic protein. Even more crucial is the fact that the obtained interactions are not able to reach the native state regardless of the starting conformation. The simulated chain of residues has first to be artificially brought to its native state (or to a very similar conformation). Then the protein may be stabilized in its fluctuations around its native state by applying the effective interaction Hamiltonian. It is clear that the limitation of the possible starting conformations implies the knowledge, *a priori*, of the native state of the protein. It follows that this method can not be applied to proteins with an unknown native state (which is eventually the aim of protein folding).

In conclusion, neither the two main techniques described above, nor other "alternative" methods, even if useful for some specific cases, have yet been brought to a satisfactory level so as to be efficient for the investigation of the whole protein folding process of any given protein.

*Protein design* is, in some sense, the opposite problem of protein folding: its aim being that of finding an amino-acid sequence that has as native state a *formerly chosen* geometrical conformation. The rationale behind this scheme is that a protein designed to interact with a particular site of a target should have a given shape, to be geometrically compatible, and suitable amino-acids on the surface to provide the right chemical properties. The solution of this puzzle would have an enormous impact especially on the development of new drugs, since nowadays, biochemical laboratories are able to synthesize and replicate any desired amino-acid sequences.

Even if protein design appears, at first glance, as the specular problem of protein folding, it nevertheless differs from it in several aspects. For instance, it follows from the Anfinsen's dogma that the relation between the *sequence space* (*i.e.* the space of all possible sequences of residues) and the *conformation space* (the space of all possible conformations) is injective in the case of protein folding: a given amino-acid chain always folds in the same three-dimensional structure. On the contrary, in the case of protein design, different primary structures can be designed for the same target conformation. Indeed, as mentioned above, only shape and surface composition are relevant for the protein function. It follows that several sequences of residues may be suitable for the same protein: each amino-acid chain folding in a different conformation with different core but nevertheless with same shape and surface composition.

For the same reasons as for the folding problem, protein design can not be tackled by all-atoms techniques. The complexity of the system has therefore to be reduced in order to make possible, at least in principle, the entire investigation of both sequence space and conformation space. Once a suitable model of the system is chosen, one has to be able to assign a cost $C$ to all possible target structures $\Gamma$ when mounting a given sequence $S$ on them. Then, $\Gamma$ is the native state of $S$ if $C(S,\Gamma) < C(S,\Gamma')$, for any structure $\Gamma' \neq \Gamma$ [15]. A suitable cost function, and the most intuitive one, could be one of the Hamiltonians in use for protein folding. However, even if $20 \times 20$ interaction matrices are useful for real

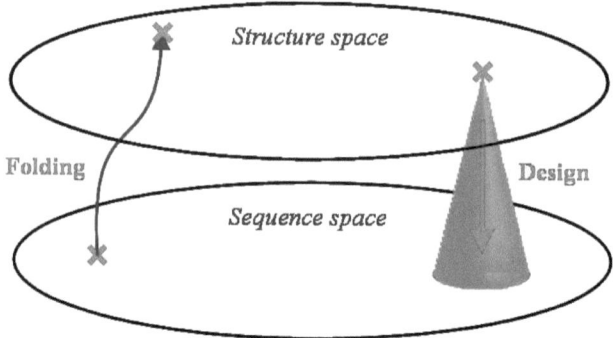

Figure 1.7: Schematic representation of the relation between sequence space and conformation space. While a given sequence folds in a unique structure, a target conformation can be reached by different residue sequences. This non-equivalence of the two processes is due to the fact that a target structure is mainly determined by its residues on the surface, whereas the amino-acids in the core are not relevant for the function of the protein.

protein design, they are too complex to address general questions. Yet, proteins can be designed with simpler Hamiltonians, for example by taking into account the *hydrophobic effect* induced by the solvent on the residues.

## 1.4 Solvent effects and hydrophobicity

Proteins *in vivo* are surrounded by an *aqueous solution*, which plays a fundamental role both in the dynamics and in the thermodynamics of folding through the *hydrophobic* properties of non-polar amino-acids [19].

*Hydrophobic substances* are defined as those that are easily soluble in non-polar solvents, such as, for example, benzene, but are badly soluble in water because of the strong polarity of its molecules. Consequently, *hydrophobicity* is commonly described as the propensity of non-polar molecules to reduce the solute-water interface: in a polar solvent, hydrophobic particles have a tendency to stick together hiding mutually their contact surface, so as to minimize the global surface exposed to the solvent.

For about half a century, hydrophobic interactions have been considered as a leading force for the folding process and a primary cause of the stability of proteins [19]. Indeed, because of the propensity of non-polar amino-acids to minimize the surface exposed to the solvent, a considerable number of proteins (in particular, all those referred to as *globular* proteins) have its native state constituted by a core made by hydrophobic amino-acids surrounded by a layer mainly composed of polar residues. However, despite the evidence and the general agreement about the fundamental role of the hydrophobic effect in protein folding, there is still a lack in the comprehension of the real mechanism underlying hydrophobicity. For long time, it was believed that aggregation of hydrophobic molecules in a polar solvent was the result of attractive interactions between them [20, 21]. On the contrary, it has more recently been shown that this effect is principally caused by strong attractive forces (the so-called *hydrogen bonds*) between isotropically-arranged water molecules, whereas the intermolecular interactions of non-polar molecules play only a minor role [22, 23]. The insertion of a particle in water causes distortions and eventually disruption of the hydrogen bonds, inducing an enthalpy loss of the system. However, in the case of ionic or strong polar molecules, bonds between the solvent and the particles can be established, which more than compensate the loss of cohesion among water molecules. Therefore, these substances are easily soluble in water and are known as *hydrophilic* compounds. On the other hand, non-polar particles are unable to form hydrogen bonds: the insertion of hydrophobic molecules in aqueous solution originates a disruption of hydrogen bonds in the solvent which is not compensated by particle-solvent interactions. Hence, the solvent tries to aggregate the solute molecules in order to minimize the number of disrupted bonds, thus inhibiting the solvation of non-polar substances.

The crucial role of the solvent in the mechanism responsible for the hydrophobic effect was already identified by Frank and Evans, while investigating the thermodynamic properties of aqueous solutions of several different kinds of hydrophobic substances [22]. Their experiments highlighted the ability of liquid water to create highly-ordered states by means of extended *hydrogen-bond networks* (Fig. 1.8) and their partial disruption when inserting particles in the solution. As for proteins, about ten years later, Kauzmann's experiments stressed the fundamental role of the solvent both in the folding and in the stabilization of proteins, emphasizing the singular properties of the physiological solvent and its indispensable role for life [19].

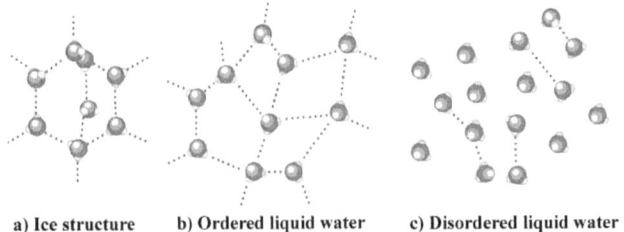

a) Ice structure     b) Ordered liquid water     c) Disordered liquid water

Figure 1.8: Schematic representation of three different states of water: crystalline ice (a), ordered liquid water (b) and disordered liquid water (c). Ordered liquid water is characterized by rather strong hydrogen bonds, which are able to create extended networks. On the other hand, the structure of disordered water is composed by weak hydrogen-bonds and many unbounded molecules. It follows that disordered states possess lower density as the ordered ones. Oxygen atoms are shown in red and hydrogen atoms in white, while hydrogen-bonds are represented by dotted red lines.

In principle, the disruption of hydrogen bonds caused by insertion of non-polar particles in aqueous solutions should be associated with an enthalpy loss of the system. On the contrary, detailed studies of the thermodynamic properties of these systems evidenced, at room temperature, a very large heat capacity for transfer of hydrophobic molecules to water [24, 25]. This phenomenon (known as *hydrophobic heat-capacity anomaly*) implies a decreasing solubility of non-polar particles while increasing the temperature close to room temperature, behavior that has been experimentally proved for small hydrocarbons.

It is generally believed that the reason for the hydrophobic heat-capacity anomaly lies in a structural change of the solvent induced by the added hydrophobic molecules [26]. Consequently, also the mechanism underlying the hydrophobic effect has to be directly related to this atypical behavior of water. Indeed, experiments evidenced, for the case of non-polar residues, the ability of water molecules to create *ice-like cages* around hydrophobic particles formed by even slightly stronger hydrogen bonds with respect to those of pure liquid water (Fig. 1.9), thus obtaining an enthalpy gain of the solution [22, 26, 27, 28]. The aggregation of hydrophobic molecules is therefore not caused by attractive interactions between solute molecules, neither by some solute-water interactions. On the contrary, it is the result of an entropy gain caused by the reduction of the local restructuring of water: the minimization of the global solute surface allows to reduce the number of highly ordered water molecules needed to form the cages. Because of the particular rearrangement around non-polar molecules exhibited by water molecules, the latter can be divided into two groups: those in the *hydration shell* (*i.e.* around non-polar particles) and therefore involved in the formation of cages, and those in the *bulk*, which are implicated in the generation of the hydrogen-bond networks.

The structure of aqueous solutions containing non-polar substances is strongly temperature dependent, since the free energy of formation of cages around hydrophobic particles is a balance between an entropy loss/gain and an enthalpy gain/loss. At temperatures significantly lower than room temperature, the composition of the solution is an homogeneous mixture, and hence no observable aggregation of hydrophobic molecules is present: the enthalpy term in the free energy is dominant, favoring the creation of cages around single molecules of the solute. On the other hand, the solubility of the solute decreases upon increasing the temperature and a transition phase occurs at a critical temperature, known as *Lower Critical Solution Temperature* (LCST) [29, 30]. This phase transition transforms the homogeneous mixture into the typical phase state present at room temperature: the enthalpy gain obtained by creation of cages is not large enough to compensate the loss of entropy induced by the local ordering of water molecules, the solute molecules collapse generating hydrophobic aggregates. Furthermore, increasing the temperature above the room temperature, the solubility of the solute, after reaching a minimum, increases again, since the *mixing* entropic term of the free energy is overcoming the enthalpic one. Eventually, a further phase transition occurs at the so-called *Upper Critical Solution Temperature* (UCST): the mixing entropic term is dominant, molecules are disordered and the system regain an homogeneous phase [31]. This behavior is pictorially shown, at a molecular level, in Fig. 1.10.

The structural rearrangement of water molecules around non-polar particles and its particular temperature dependence play a fundamental role both for stability and for denaturation of proteins. It has long been known that proteins are very susceptible of the

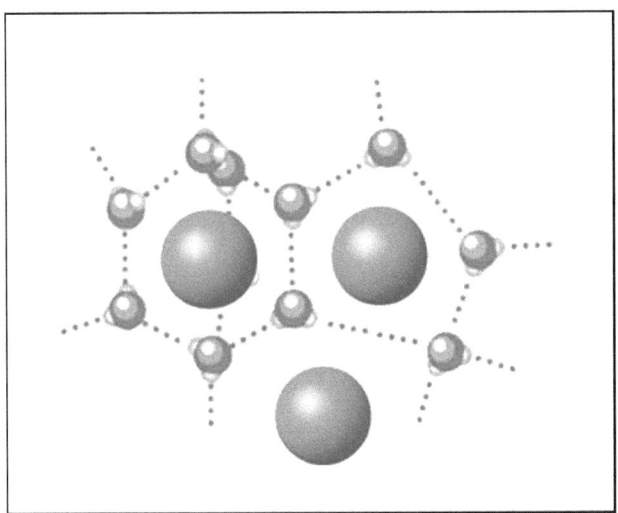

Figure 1.9: Schematic representation of water structures around non-polar molecules. Water molecules rearrange themselves in highly ordered *ice-lice cages* surrounding the solute particles. These structures favor the creation of strong hydrogen-bonds implying an enthalpy reduction in the system.

milieu modifications (such as for example pH changes) and, in particular, their behavior is strongly temperature dependent: near room temperature proteins are stable, but become rapidly unstable away from it. Indeed, by heating proteins in solution, a critical temperature is quickly reached, where a first order phase transition occurs, known as *heat denaturation*: under entropic pressure, proteins denaturate, destroying their native state and fluctuating among various open conformations.

While heat denaturation (also referred to as *warm* denaturation) has already been discovered at the dawn of the biochemistry era, only recent experiments revealed that the free energy difference $\Delta G_{DN}$ between denaturated and native conformations of proteins has a parabolic shape, with a maximum at temperatures of the order of 15-25 °C (Fig. 1.11). It follows that, besides the well-known heat denaturation, a further phase transition occurs while decreasing the temperature below room temperature. Indeed, in the last fifteen years there has been a growing body of evidence for the so-called *cold* denaturation: below room temperature, proteins become less and less stable and eventually denaturate if brought past the phase transition temperature [27, 32, 33, 34, 35, 36, 37, 38, 39, 40]. The reason of the time gap between the discovery of warm denaturation and cold denaturation resides in the fact that the latter may occur for temperatures below the freezing point, and therefore can only be induced by application of special techniques, such as supercooling. However, this phenomenon is an intrinsic property of proteins and should therefore be reproducible by any protein modelization.

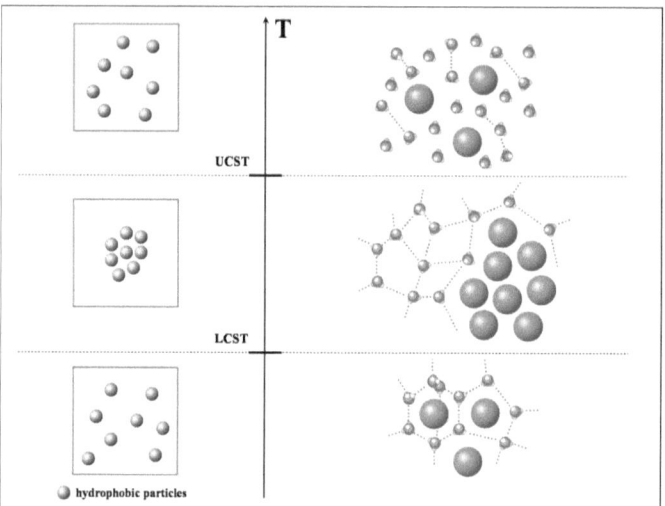

Figure 1.10: Schematic representation (both at a macroscopic (left) and at a molecular level (right)) of the solubility of non-polar molecules in water and its dependence on temperature. Between the Lower Critical Solution Temperature (LCST) and the Upper Critical Solution Temperature (UCST) the phase of the system corresponds to the well-known inhomogeneous solution: aggregates of solute particles are present, because non-polar molecules, under hydrophobic pressure, tend to minimize the global exposed surface to the solvent. On the other hand, below the LCST, respectively above the UCST, the phase of the system is an homogeneous solution.

It is clear that the behavior or proteins, showing a compact conformation (possessing an hydrophobic core) in a limited range of temperatures and denaturated states above and below this region, is directly related with the particular rearrangement of water molecules in the presence of hydrophobic particles (in the special case, the non-polar residues of the amino-acid chain). This relation becomes even more evident while comparing Fig. 1.11, where the behavior of proteins is pictorially shown, with Fig. 1.10, representing the behavior of free non-polar molecules in water. It follows that the hydrophobic effect plays a fundamental role both for folding and for stability of proteins and has therefore to be taken into account while investigating dynamical and thermodynamical properties of proteins.

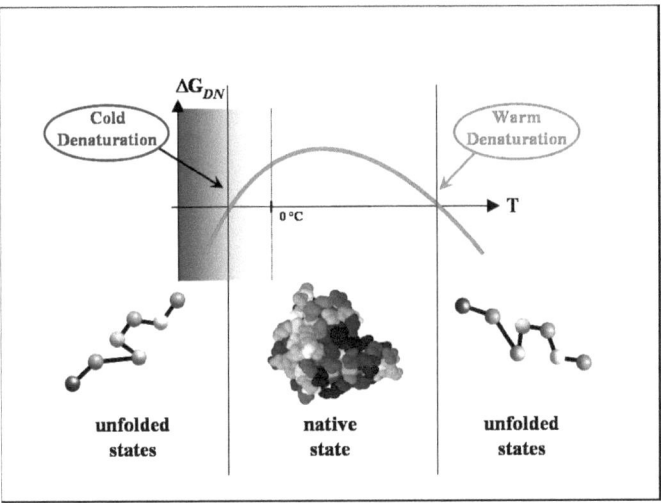

Figure 1.11: Schematic representation of the free energy difference $\Delta G_{DN}$ between denaturated and native conformations of proteins and its dependence on temperature. Because of its parabolic shape, two phase transitions occur: proteins are only stable in a limited range of temperatures. Either by heating protein solutions above the *warm denaturation* temperature or by cooling it below the *cold denaturation* temperature, the rearrangement of water molecules causes the disruption of the protein native states and consequently denaturation occurs.

## 1.5 Cosolvents

Studies of protein folding often involve the use of chemical denaturants, such as urea and guanidinium hydrochloride, that increase the solubility of hydrophobic amino-acid groups and consequently decrease the stability of the native states [41]. Despite their widespread use, however, the mechanism of their action is not well understood.

Only lately there has been growing interest in cosolvents, and, in particular, recent experiments focused on the properties of the class of cosolvents (such as urea) known as *chaotropic substances*. It turned out that, even if cosolvent molecules may bind directly to proteins, their effect appears to be mainly related to a modification of the solute properties. Indeed, it is generally believed that the capability of chaotropic substances to increase the solubility of hydrophobic particles in aqueous solutions resides primarily in their ability to decrease the order of the water structure ('chaotrope' = disorder maker) [42, 43, 44, 45, 46, 47, 48, 49, 50, 51, 52]. The perturbation in the hydrogen bond networks of water molecules, caused by the addition of chaotrope molecules, results in a weaking of the aggregation of the solute particles and can enhance the solubility of hydrophobic molecules by several orders of magnitude. Eventually, for high cosolvent concentrations, denaturation of proteins occurs.

It has been shown that chaotrope molecules are not homogeneously distributed when inserted in aqueous solutions: close to non-polar particles the local concentration of cosolvent is larger than the average cosolvent concentration in the solution (Fig. 1.12) [52]. This effect, known as *preferential binding*, implies a smaller number of water molecules in

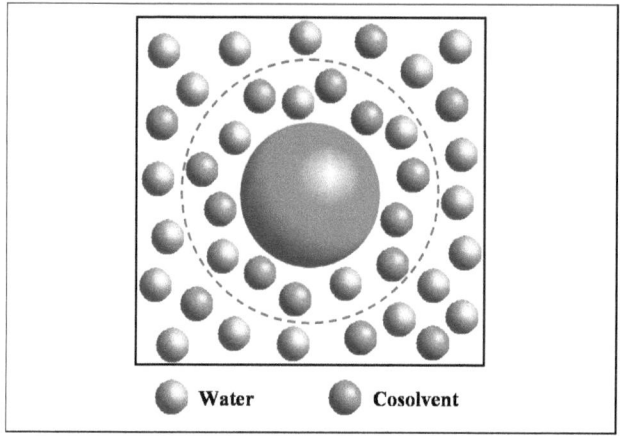

Figure 1.12: Schematic representation of the *preferential binding* effect. Molecules of chaotropic substances, such as urea, while diluted in aqueous solutions, have a tendency to stay close to non-polar molecules. It follows that the local concentration of cosolute in hydration shells is higher than the average concentration in the solution.

the hydration shells and consequently a weakening of the hydrophobic effect. Indeed, the reduction of the number of water molecules in contact with non-polar residues leads to a weaker net interaction between solute and solvent, such that it becomes more favorable to increase the global solute surface in contact with the solvent. Therefore, the latter tends to disrupt the solute aggregations in order to maximize their exposed surface. It follows that the addition of chaotropic substances in protein solutions destabilizes and eventually denatures proteins.

In this study, by implementing a cosolvent into our original lattice model, we show that the disordering effect of chaotropic substances on the water structure, without any direct interaction with the solute, is sufficient to exhibit both known properties, *i.e.* preferential binding and denaturation of proteins.

## 1.6 Overview

In this Ph.D. thesis, we investigate the solvent effects and the resulting *hydrophobic effect* on proteins, mainly dealing with a simple lattice model of proteins: the HPW model [53, 54]. Chapt.2 introduces a detailed description of the models used in this work, namely the HP model (*implicit* solvent) [55, 56] and the HPW model (*explicit* solvent) with the related double bimodal model MLG [57, 58] employed to express the solvent. In Chapt.3 we approach the protein design problem by means of the HPW model and we investigate some statistical and thermodynamic properties of the designed sequences, such as, for example, *stability* and *denaturation* of proteins as a function of the temperature $T$. Chapt.4 is devoted to cosolvent effects on the stability of proteins. We modify the MLG model in order to introduce chaotrope agents in the HPW model and we investigate the behavior of proteins for different cosolvent concentrations. In particular, we are interested in the changes of the free energy difference $\Delta G_{DN}$ between the native state and the unfolded states induced by chaotrope agents. Finally, in Chapt.5, we introduce a common method to extract effective amino-acid interactions on real proteins: the perceptron algorithm. We apply this method to proteins designed with the HPW model in order to study the possibility of reproducing the solvent effects on proteins by means of effective residue interactions. Chapt.6 contains a summary and our conclusions.

# Chapter 2

# The models

*All-atom* techniques are far too complex to address general questions about proteins. Therefore, models have to be introduced in order to reduce the complexity at a manageable level and, at the same time, keeping the essential features of the system composed both by the protein and by the solvent.

Lattice models are commonly used in a wide range of physical problems. It is therefore not surprising that this approximation technique was also introduced in protein research. The main advantage of dealing with lattice models of proteins is the strong reduction of the number of degrees of freedom for the system so that, at least in principle, the investigation of the whole conformational space is possible. In particular, in the present work we deal with two models both being applied on a *two-dimensional square* lattice where amino-acid sequences are described as *self-avoiding walks* (SAW). Indeed, since real proteins are subject to excluded volume, the choice of self-avoidance for the modeled chains is a logical consequence in order to keep this property in the model. As for the choice of the two-dimensional space, even if it is a difficult task to address the effect of dimensionality, two-dimensional models offer at least a considerable advantage over three-dimensional ones. Indeed, a fundamental physical variable of the polymer collapse theory is the ratio between the monomers at the surface and those present in the core. For instance, a realistic ratio value for small globular proteins composed by about 100 monomers, is 1/4. While working on a three-dimensional square lattice, it is easy to see that the smallest conformation containing a core where no residues are exposed to the solvent is composed of 27 monomers, *i.e.* the $3 \times 3 \times 3$ cubic structure with one residue in the core and 25 monomers at the surface. On the contrary, in the two-dimensional case, 9 residues are enough to create a compact conformation (*i.e.* the $3 \times 3$ square structure), while 16 monomers allow the creation of a $4 \times 4$ square structure with already 4 residues forming the core and consequently not exposed to the solvent. While working with exhaustive enumeration, the length of the chains is strongly limited, since the number of possible conformations increases exponentially with the length of the chain. It follows that regarding the inside/outside ratio, when dealing with short chains, the two-dimensional lattice is a better approximation of the real case.

Since there are 20 naturally occurring amino-acids (each of them with its own sterical and chemical properties), modelization of proteins has to take into account, at least in principle, all different properties of the residues. However, while working with simplified models, there is usually no need to deal with an amino-acid alphabet composed by 20 different residues. Indeed, rather than assigning unique properties to each amino-acid, the alphabet can be reduced by dividing residues into groups: elements of the same group share the same properties. For instance, while working on a lattice, the difference in sterical properties can not be taken into account, since any amino-acid is able to occupy any site of the lattice, regardless to its size. Therefore, for simple lattice models only chemical properties are meaningful.

It is nowadays well-established (since confirmed both by experimental and by theoretical works) that the driving force of folding of globular proteins is the hydrophobic effect [19, 27, 59]: hydrophobic amino-acids try to hide their surface from the solvent, creating a hydrophobic core surrounded by polar residues. It follows that this force has to be implemented in any protein model that tackles with problems related either to the folding or to the stabilization of proteins. A commonly used method, introduced to exploit this property, is to reduce the amino-acid alphabet to only two species: a given residue is either *hydrophobic* (H) either *polar* (P) [55, 56]. This simplification is apparently a very strong assumption since it neglects any further chemical property (such as, for example, differences both in charge sign and in strength). However, it allows to focus on the interactions created by the hydrophobic effect, rather than the more subtle specific types of interactions, belonging to the different kinds of residues. In addition, the reduction of the number of different amino-acids, from the originally twenty to only two, shrinks drastically the sequence space, allowing the whole investigation of the latter.

Since the hydrophobic effect responsible for the aggregation of non-polar residues is induced by the solvent, the latter has to be implemented so as reproducing the correct behavior of the amino-acid chain. However, because of the complex (and still not completely understood) behavior of water, and moreover, because of the huge number of involved molecules, also the description of the solvent has to be drastically simplified. One common technique is to implement the solvent in an *implicit* form, *i.e.* the solvent it not "physically" present in the model but, on the contrary, its effects are described by effective forces acting directly on the amino-acids of the protein. However, it is known that, even if the full description of the solvent can not be applied, it is nevertheless possible to deal with a simplified representation of it, allowing the introduction of the solvent in the system in an *explicit* form. This method has clearly the big advantage over the implicit technique to avoid the introduction of some kind of artificial effective interactions between amino-acids. In particular, it has recently been shown that, by a smart choice of the basic properties of water, an explicit description of the solvent is possible even in a very simple lattice model.

In the next two sections, we describe the two models used in this book: the HP model [55, 56], with an *implicit* representation of the solvent and the HPW model [53, 54], characterized by a *(semi)explicit* solvent description. However, we recall that this study is focused on the latter, while the HP model is mainly used for comparison of the results obtained with the HPW model.

## 2.1 The HP model

This implicit solvent model, introduced by K. F. Lau and K. A. Dill in 1989 [55], was the first one dealing with proteins on a lattice and with a reduced number of different amino-acids. In particular, it was the first attempt to thoroughly investigate the influence of the hydrophobic effect (induced by the solvent) on the protein stability by means of a simple lattice model.

In order to focus on solvent effects, the HP model handles with only two kind of amino-acids: *hydrophobic* (H) residues and *polar* (P) residues. It follows that proteins are binary chains arranged on a square lattice with the only restriction that no crossings are possible along the chain. In other words, all possible conformations of a chain of a given length $N$ are simply all possible *self-avoiding walks* (SAW) obtained with $N$ steps.

Once a given sequence of residues $S$ is mounted on a particular conformation $\Gamma$, contacts between neighbor residues can be divided into two groups: the *connected contacts* are those between residues $i$ and $i+1$ (where $i$ is the index of the residue position in the chain), trivially associated by the intrinsic connectivity of the chain (the peptide bonds); while the interesting contacts, since depending of the given conformation $\Gamma$, are the so-called *topological contacts* (or *non-covalent* bonds), formed by neighbor residues that are not sequentially positioned along the chain.

Since the hydrophobic effect induces an aggregation of non-polar particles (if the latter are introduced in an aqueous solution), in the HP model the solvent effects are captured by an attractive interaction $\epsilon(\epsilon < 0)$ between hydrophobic residues connected by a topological contact. It follows that, with this implicit solvent description, the partition function of a chain of fixed length $N$ becomes:

$$Z = \sum_{m=0}^{s} g(m) e^{\beta(s-m)\epsilon} \qquad (2.1)$$

where $m = 0, 1, 2, \ldots, s$ is the number of HH contacts, *i.e* the number of topological contacts between hydrophobic residues, and $g(m)$ is the degeneracy, *i.e.* the number of chain conformations with exactly $m$ HH topological contacts ($\beta^{-1} = k_B T$). Starting from the partition function it is easy to compute all relevant thermodynamic properties, such as the average energy $<E>$, the specific heat $C_v$ and more specific quantities, for instance the average compactness $<\rho>$ and the closely related average gyration radius $<R_G>$.

In the last ten years, this model has been thoroughly investigated, computing and analyzing all relevant properties. It turned out, that despite the strong simplifications, it provided many results in agreement with properties of real proteins. In particular, this model reproduces a fundamental property of globular proteins, *i.e.* the presence of a hydrophobic core in the native state. At low temperature, the chain of residues is folded in the compact conformation possessing the maximal number of topologically possible HH contacts. Indeed, from the partition function of Eq. (2.1) it is easy to see that for $T \to 0$ the dominating term in the partition function is the one with the maximum number of HH contacts. It follows that, at low temperatures, the chain collapses

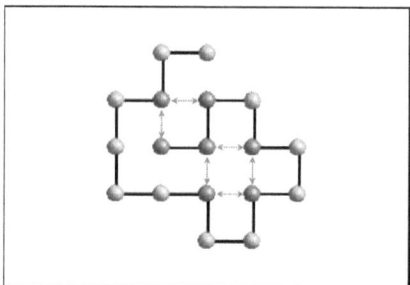

Figure 2.1: The two-dimensional HP model: a typical compact conformation for the sequence *HHHPHPPHPPHPPPPHPP*. The green spheres represent the polar amino-acids, while the hydrophobic ones are represented by red spheres. The effective interactions $\epsilon$ between H amino-acids are pictorially shown with red arrows. The core of the protein is composed by hydrophobic amino-acids and is surrounded by polar ones.

in a compact conformation (referred to as *native state*) where the hydrophobic amino-acids are mainly present in the core of the protein: aggregating the hydrophobic residues in the core is the only possible method to maximize the number of HH contacts (Fig. 2.1).

While at low temperatures the conformation of the chain of residues is the native state, by rising the temperature *heat* denaturation occurs. Indeed, as for real proteins, by heating the system, a first order phase transition occurs and the protein denaturates: for high temperatures, the entropic gain of the unfolded conformations of the chain overcomes the enthalpy gain obtained by the larger number of HH interactions of the native state (Fig. 2.2).

These two important results of the HP model, namely the presence of a hydrophobic core in the native state and the occurrence of heat denaturations, are very relevant for protein research. Indeed, with this simple lattice model it has been confirmed that the hydrophobic effect plays a crucial role for the stability of proteins.

In the past years, this model (nowadays referred to as *standard* HP model) was improved by adding further properties. For instance including the HP and the PP contacts, *i.e.* contacts between hydrophobic and polar amino-acids, respectively only between polar residues. Obviously, since the hydrophobic effect mainly acts on the hydrophobic residues, the choice of the interaction potentials ( *i.e.* the energies of the different possible couples of residues) has to satisfy the following inequalities: $\epsilon_{HH} < \epsilon_{HP}(= \epsilon_{PH}) < \epsilon_{PP}$. Furthermore, once the values are fixed, it is possible to compute the partition function of the system, since the energy of a given sequence $S$ mounted on a conformation $\Gamma$ becomes:

$$H(S,\Gamma) = \sum_{i<j} \epsilon^{nn}_{p_i p_j} \Delta_{nn}(\vec{r}_i, \vec{r}_j), \qquad (2.2)$$

where $\epsilon^{nn}_{p_i p_j}$ are the interaction potentials ($p_i$ = H, P) and $\Delta_{nn}(\vec{r}_i, \vec{r}_j)$ is a contact matrix

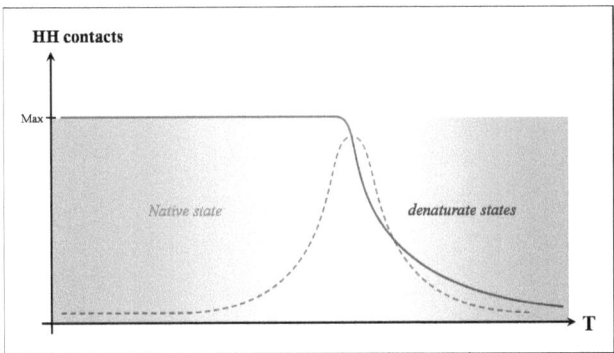

Figure 2.2: Schematic representation of the temperature dependence of a protein in the HP model: the number of HH contacts are represented by the green line while the dashed red line represents the specific heat of the system. At low temperature, the protein is in its native state, characterized by a maximal number of HH contacts. By rising the temperature, *heat* denaturation occurs: the native state is destroyed and the chain fluctuates among open conformations.

defined as:

$$\Delta_{nn}(\vec{r}_i, \vec{r}_j) = \begin{cases} 1 & \text{if } i,j \text{ are nearest-neighbors} \\ 0 & \text{otherwise.} \end{cases} \quad (2.3)$$

Moreover, it was also possible to extend the model to a three-dimensional square lattice (Fig. 2.3). Indeed, even if the whole investigation of both the sequence space and the conformation space is only feasible for short chains (typically chains with length $L \leq 16$), investigation is nevertheless possible also for longer chains, as needed for the three-dimensional case. For instance, it is possible to reduce the conformational space to a subset $\{\Gamma\}_c$ containing only the compact conformations, since it is known that the native states are compact conformations. This reduction may therefore allow the use of exhaustive enumeration also for sequences considerably longer than $L = 16$. A further common technique is to focus on only a small subset $\{S\}_i$ of sequences, usually possessing some particular properties. In addition, as for the case of the three-dimensional square lattice, Monte-Carlo simulations are applied and combined with the reduction of the number of investigated conformations, so that it was possible to obtain valuable results for chains of length up to $L = 46$.

All these further models were also thoroughly investigated and, beside the two main results (the hydrophobic core of the native state and heat denaturation) many other results were obtained. For instance, secondary structures in the native states have been widely investigated. It turned out that secondary structures of chain molecules are a natural consequence of compactness and not, as first presumed, of the complexity of the amino-acid alphabet and the related sterical differences, nor of subtle chemical forces.

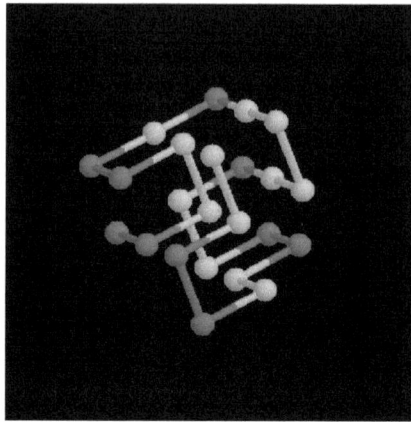

Figure 2.3: Example of a protein structure obtained with the three-dimensional HP model for a sequence of length $L = 23$. Gray spheres represent polar residues, while hydrophobic amino-acids are represented by the yellow ones.

## 2.2 The HPW model

Even if the HP model provided many interesting results, the effective accuracy of describing solvent effects by the introduction of interactions between hydrophobic amino-acids is still an open issue. Therefore, few years ago, another model was proposed by P. De Los Rios and G. Caldarelli [53, 54], where the solvent is (semi)explicitly taken into account. Their model borrows the two main simplifications of the standard HP model: proteins are heteropolymers described as self-avoiding walks on a two-dimensional lattice, while the amino-acids are encoded in only two species, *i.e.* they are either hydrophobic (H) either polar (P). However, as mentioned above, the difference between the two models resides in the description of the solvent effects. Indeed, in this new model, the hydrophobic interaction is not expressed by an effective interaction between hydrophobic residues, but rather by the solvent behavior. For this purpose, the model has, in some sense, to put proteins back into water: once a given sequence $S$ is mounted on a conformation $\Gamma$, every site of the lattice not occupied by a monomer of the chain is filled by groups of water molecules (introducing some coarse graining in the model). Because of the similarity with the HP model and of the presence of water, this model is referred to as HPW (HP+Water) model. Finally, a simple model which describes the essential physics of the system, now composed by the amino-acids and by the solvent, is required. Regarding the solvent, the minimal model describing the water behavior as an aqueous solvent is the double bimodal description of Muller, Lee and Graziano (MLG), where only four different states are accessible by any single water molecule [57, 58].

It has long been known that liquid water is able to create highly-ordered states by means of extended hydrogen-bond networks. Groups (or *clusters*) of water molecules can therefore be characterized by the number of their hydrogen-bonds. The MLG model simplifies this property by dividing water molecules into only two populations: clusters of molecules in *ordered* states are those possessing a large number of hydrogen-bonds, while clusters where molecules are only weakly interacting are in one of the remaining possible *disordered* states.

Additionally, since water molecules are able to create ordered structures (in particular by creating ice-like cages) around non-polar particles, the MLG model also divides water molecules into two groups depending on their positions. Sites close to a hydrophobic particle (and therefore affected by its presence) are referred to as *shell* sites (and belong to the so-called *hydration shell*). In contrast, all sites not belonging to any hydration shell are referred to as *bulk* sites.

Since also clusters of water molecules in the hydration shells can be characterized by their number of hydrogen-bonds, in the MLG model molecules belonging to shell sites are also divided into two groups: those being in ordered states and those present in disordered states. In summary, with this bimodal description of the solvent, water molecules are present in four different states, depending on their position (shell sites ($s$) or bulk sites ($b$)) and on the number of their hydrogen bonds (ordered states ($o$) or disordered states ($d$)) (see Fig. 2.4).

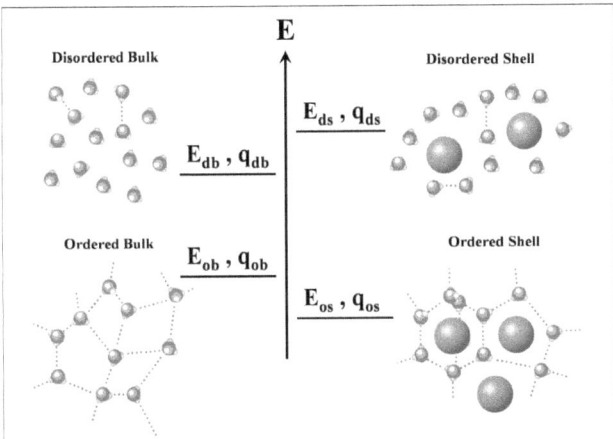

Figure 2.4: Energy levels and degeneracies in the MLG model for water. On the right of the central arrow the values of the shell sites, *i.e.* close to a non-polar particle, are represented, while the bulk sites are those on the left. The energy levels are arranged so that $E_{ds} > E_{db} > E_{ob} > E_{os}$, while the degeneracy levels follow the inequalities: $q_{ds} > q_{db} > q_{ob} > q_{os}$ (see text). The different states of water at a molecular level are schematically illustrated.

The determination of the correct energy hierarchy, respectively degeneracy hierarchy, among the different states is crucial in order to reproduce the correct behavior of the solvent. However, these values are easily estimated by means of some simple qualitative arguments. For instance, because of the larger number of hydrogen-bonds, ordered states (for both shell and bulk water) are energetically favorable with respect to disordered states, hence $E_{ds} > E_{os}$ and $E_{db} > E_{ob}$. In addition, water molecules creating hydrogen-bonds have a higher degree of order and thus fewer rotational degrees of freedom than those unbounded. It follows that the disordered states outnumber the ordered ones: $q_{ds} > q_{os}$ and $q_{db} > q_{ob}$.

While the above arguments hold separately for bulk and shell sites, relations between both energy and degeneracy of the two different sites are provided by further considerations. Indeed, on one hand, it has been shown that ordered structures in the hydration shell are on average stronger than those in bulk water, *i.e.* $E_{ob} > E_{os}$. On the other hand, for steric reasons, fewer hydrogen-bonded water configurations are possible around a non-polar molecule. In fact, since the latter is unable to form hydrogen-bonds, all conformations for which water molecules are pointing toward the hydrophobic particle are not considered as ordered states (because of the absence of the hydrogen-bond). On the contrary, if the hydrophobic particle is replaced by water molecules, additional ordered states are possible, since the molecules may form further radial hydrogen-bonds with those located in the center of the cluster, hence $q_{ob} > q_{os}$.

Finally, regarding disordered states, the average energy $E_{ds}$ of clusters of disordered molecules in a hydration shell decreases when the non-polar particle is replaced by water. Indeed, as explained above, the presence of water molecules in the center of the cluster allows the creation of further radial hydrogen-bonds. It follows that, because of the larger number of hydrogen-bonds, the average energy of clusters of disordered molecules is lower in the bulk than in the shell, *i.e.* $E_{ds} > E_{db}$. As for the degeneracy, fewer disordered states are present in the bulk than in the hydration shells. Indeed, in the bulk, all orientations of molecules which allow to form hydrogen-bonds with those in the center, are considered as ordered states. On the contrary, by inserting a non-polar particle, the radial hydrogen-bonds of those orientations are broken and therefore considered as disordered states. It follows, that the disordered states in the shell outnumber the one in the bulk, so that $q_{ds} > q_{db}$.

In summary, all the above considerations lead to the following relations between the different values: $q_{ds} > q_{db} > q_{ob} > q_{os}$ for the degeneracies and $E_{ds} > E_{db} > E_{ob} > E_{os}$ for the energies. In addition, it is important to mention that, even if the arguments used above are mainly qualitative, such a double bimodal description of water (better represented pictorially in Fig. 2.4) has been successfully used to fit experiments of solvation and re-derived by both simple and realistic models of water [58, 60].

Moreover, regarding the number of parameters to adjust, it is important to stress that the *effective* number of parameters in this model is four and it has also been shown that the same qualitative behavior of the solvent can be reproduced even simplifying this model reducing the number of parameters to only three [53]. Regarding the four effective parameters: the origin of the energies has been chosen so that pairs of levels are symmet-

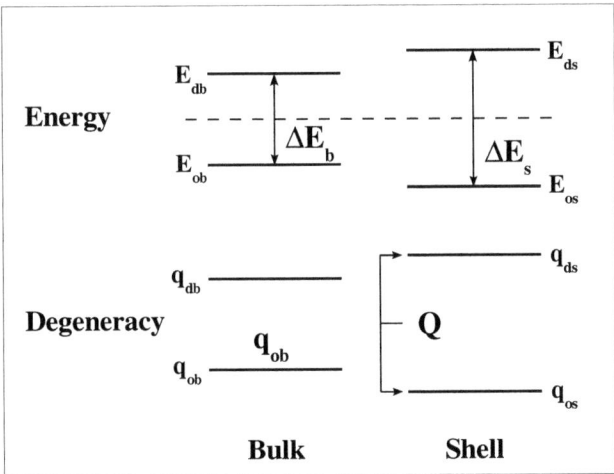

Figure 2.5: The effective parameters of the MLG model. The origin of the energies is chosen so that pairs of levels are symmetrically spaced with rapport to it. This choice implies fitting of two energy parameters, *i.e.* the energy gap between ordered and disordered states, both in the shell ($\Delta E_s$) and in the bulk ($\Delta E_b$) sites. Regarding degeneracies, once one of them has been arbitrarily set to 1, only two further parameters are needed in order to fix all four: the first one being the overall degeneracy $Q(= q_{os} + q_{ds} = q_{ob} + q_{db})$ of bulk and shell sites, whereas the second one is their bulk partition, which is trivially defined, fixing the number of ordered states in the bulk $q_{ob}$.

rically spaced with rapport to it and the degeneracy of the ordered shell sites $q_{os}$ has been set to 1. Finally, since the overall degeneracies $Q$ of bulk and shell sites are equal, only their bulk partition is an extra parameter (see Fig. 2.5).

In the HPW model, the MLG model is used to describe the behavior of groups of water molecules. Indeed, as introduced above, in the HPW model every site of the lattice not occupied by an amino-acid is taken by a group of water. Moreover, in order to apply the bimodal description of water, the sites of the lattice are divided into two groups: each site of the lattice nearest-neighbor of at least a hydrophobic monomer is considered as shell site, while the remaining are regarded as bulk sites.

The use of a coarse-grain approximation, by considering a group of water molecules as a single unit, is justified since the size of hydrophobic amino-acids is relatively large compared to the size of a single water molecule. In fact, if the considered non-polar particles are large with respect to water molecules, the latter are not able to form a complete cage around the particle but rather only fractions of cages. In this case, the coarse-grain approximation considers, in some sense, these partial cages as a single unit.

As for the amino-acid properties, no interactions between residues are introduced in the model, since it is focused on the hydrophobic effect induced by the solvent. In fact, even if in a more realistic model chemical properties of the amino-acids have to be taken into account, dealing with only two kinds of amino-acids do not allow the introduction of some specific interactions, such as, for example, attractions between amino-acids with opposed polarity. It follows that in this model the energy of the system only depends on the properties of the solvent molecules.

The water degrees of freedom are represented by Potts-like variables $\sigma$ [61] that take $q_{os} + q_{ds}$ values for shell sites and respectively $q_{ob} + q_{db}$ if the group of water molecules is in one of the possible bulk sites. Therefore, for a given protein of $L$ amino-acids, with the sequence $S = a_1, a_2, \ldots, a_L$ ($a_i = P$ or $H$), the energy of the protein is given by the Potts-like Hamiltonian

$$H[\{\sigma_i\}, \{\sigma_j\}] = \sum_{i \in Shell} \left( E_{os} \tilde{\delta}_{i,os} + E_{ds}(1 - \tilde{\delta}_{i,os}) \right) + \sum_{j \in Bulk} \left( E_{ob} \tilde{\delta}_{j,ob} + E_{db}(1 - \tilde{\delta}_{j,ob}) \right) \qquad (2.4)$$

where the first sum is over water sites that are nearest-neighbors of some hydrophobic amino-acid, while the second sum is over all bulk sites; $\tilde{\delta}_{i,os} = 1$ if $\sigma_i = 0, ..., q_{os} - 1$, 0 otherwise, and analogously for $\tilde{\delta}_{i,ob}$; the $E_{ij}$ represent the different energies described above.

The Hamiltonian of Eq. (2.4) allows to compute the partition function of the system. Indeed, once the sequence $S$ of amino-acids is fixed, the partition function $Z(S)$ of the whole system can be written as the sum over all "partial" partition functions $Z(S, \Gamma)$, *i.e.* the partition functions (which are, in some sense, considered as *generalized* Boltzmann weights) associated to any single conformation $\Gamma$:

$$Z(S) = \sum_{\Gamma} Z(S, \Gamma) \qquad (2.5)$$

For a given conformation the number of shell sites, respectively bulk sites, are fixed. The canonical partition function is therefore

$$Z(S, \Gamma) = \sum_{\{\sigma_i\}, \{\sigma_j\}} e^{-\beta H[\{\sigma_i\}, \{\sigma_j\}]} \qquad (2.6)$$

where the sum is performed over the Boltzmann weights of all possible states in the shell ($\{\sigma_i\}$) and in the bulk ($\{\sigma_j\}$) of the water molecules ($\beta^{-1} = k_B T$). Performing that sum rises to the following result

$$Z(S, \Gamma) = \left( q_{ob} e^{-\beta E_{ob}} + q_{db} e^{-\beta E_{db}} \right)^{n_b(\Gamma)} \cdot \left( q_{os} e^{-\beta E_{os}} + q_{ds} e^{-\beta E_{ds}} \right)^{n_s(\Gamma)} \qquad (2.7)$$

where $n_s(\Gamma)$ is the number of water sites nearest-neighbors of some hydrophobic residue, while $n_b(\Gamma)$ is the total number of the remaining bulk water sites.

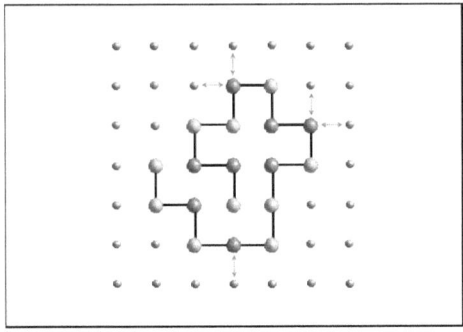

Figure 2.6: The two-dimensional HPW model: a typical compact conformation for the sequence *HHHPHPPHPPPHPPPPHPP*. The green spheres represent the polar amino-acids, while the hydrophobic ones are represented by red spheres. The shell water sites are pictorially shown with red arrows connecting the site and the related hydrophobic residue. In this model polar amino-acids are considered as bulk water sites and therefore no distinction is present between water sites close to another water site or close to a polar residue.

From the partition function we can calculate all the thermodynamic quantities. In particular, among the relevant statistical features there are the specific heat $C_v$ and the average number of shell sites $<n_s>$. These quantities were computed in former works for different sequences of length up to $L = 17$ by exact enumeration of the 2 155 667 different conformations [53, 54]. These simulations showed that the specific heat $C_v$ and the thermal average $<n_s>$ as a function of $T$ point to two different denaturations, since the $C_v$ shape contains two well defined peaks (see Fig. 2.7). Between the two peaks proteins have few contacts with water (they are in a compact state), and the most probable conformation is the one with the minimum number of shell sites $n_s(\Gamma)$, since the hydrophobic amino-acids are hidden in the core of the protein (as for real globular proteins). On the other hand, above and below the two temperatures proteins swell and the number of water-protein contacts increases.

It is known that, for real proteins, the free energy difference between unfolded states and the native state is positive below the warm denaturation temperature $T_w$, ensuring the stability of the native state; yet, this difference has a maximum around $20^o C$ for the "average" protein, and then decreases for temperatures $T < 20^o C$. Either choosing suitable proteins, or, more generally, by supercooling or by applying a pressure, it is even possible to see the *cold* denaturation of proteins in liquid water at a temperature $T_c$ that is, usually, below $0^o C$ [32]. This behavior is correctly reproduced by the proteins of the HPW model, since both warm and cold denaturation are present (represented by the two peaks of the specific heat). On the contrary, it is important to stress that this phenomenology is not reproduced by most Hamiltonians (as for example the one of the HP model) used in protein folding, since their native state is the $T = 0$ ground state (GS) of the model.

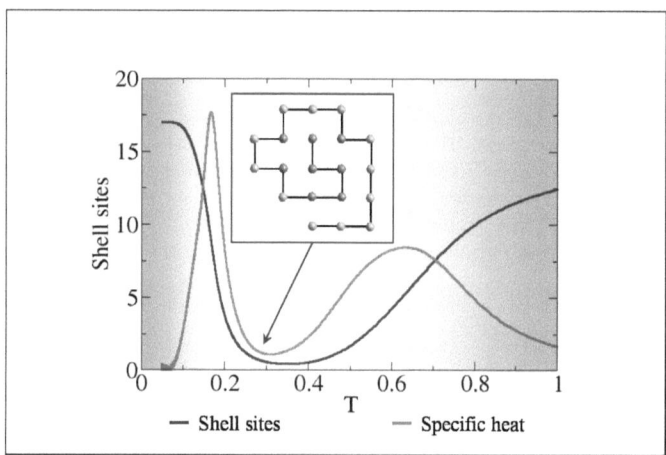

Figure 2.7: The HPW model on a two-dimensional lattice: typical behavior of the specific heat $C_v$ and the average number of shell sites $<n_s>$ of a protein. The two peaks in the specific heat $C_v$ reveal the presence of two phase transitions, namely the *cold* and the *warm* denaturation. Between the two peaks the protein is in its *native state*, *i.e.* the compact conformation possessing the minimum number of shell sites $<n_s>$: the hydrophobic amino-acids, represented by red spheres, are hidden in the core of the protein and surrounded by polar residues (green spheres) (see inset). By rising the temperature, *warm* (also referred to as *heat*) denaturation occurs because of a gain of the mixing entropy. On the other hand, by lowering the temperature, *cold* denaturation occurs, the latter being induced by an enthalpic gain given by the creation of ordered structures of the water molecules around the hydrophobic amino-acids.

# Chapter 3

# Protein Design

As explained in the previous chapters, the main goal of this study is the thorough investigation of the properties of the proteins in the framework of the HPW model. In particular, in this chapter we begin by tackling the problem of *protein design*.

The idea underlying protein design is to choose a particular geometrical conformation (referred to as *target* structure) possessing specific sterical and chemical properties on its surface, so as being able to interact with other molecules. It is important to stress that, since the target structure does require a specific shape and a specific surface composition but, on the contrary, does not require a well-defined core, the needed correct chain of residues is not unique. In other words, different sequences of amino-acids are able to fold into the same given conformation.

## 3.1 *Good* sequences are needed

A fundamental property of a sequence, in order to correctly fold in a predefined structure, is to possess a unique state with the lowest energy (conformation referred to as *native state*). Indeed, the uniqueness of the native state is a requirement to ensure that the folding to the correct conformation is not hindered by the competition of different, thermodynamically equivalent, states. While real proteins satisfy this property, since evolved through selective pressure, random sequences of amino-acids do not trivially fulfill this condition. It is therefore important to be able to individuate which sequences possess a unique native state (henceforth referred to as *good* sequences) among the whole set of virtually possible chain of residues.

Regardless the kind of used model, in order to determine which sequences can be considered as good sequences, one need to introduce a *cost function* $C(S, \Gamma)$ so that a given sequence $S$ has its native state on a target structure $\Gamma$ if $C(S, \Gamma) < C(S, \Gamma')$ for any structure $\Gamma' \neq \Gamma$. These cost functions are usually defined by one of the Hamiltonians used in protein folding.

In the HPW model, for a given sequence $S$, the partition function of any conformation

Γ is given by

$$Z(S,\Gamma) = \left(q_{ob}e^{-\beta E_{ob}} + q_{db}e^{-\beta E_{db}}\right)^{n_b(\Gamma)} \times \left(q_{os}e^{-\beta E_{os}} + q_{ds}e^{-\beta E_{ds}}\right)^{n_s(\Gamma)} \quad (3.1)$$

where $n_s(\Gamma)$ is the number of shell sites (*i.e.* those nearest neighbors of some hydrophobic amino-acids), while $n_b(\Gamma)$ is the number of the bulk water sites.

Therefore, we introduce the following value as cost function: for a given sequence $S$ mounted on a structure $\Gamma$ its cost function is the partial free energy of the conformation $\Gamma$

$$C(S,\Gamma) = -Tk_B \ln\left[\frac{Z(S,\Gamma)}{\sum_{\Gamma'} Z(S,\Gamma')}\right] \quad (3.2)$$

where $Z(S,\Gamma)$ is the partial partition function given by Eq. (3.1), whereas the sum in the denominator of the fraction is performed over all possible conformations $\Gamma'$ (*i.e.* the value of the total partition function of the system).

In addition, since the size of the lattice and the length of a sequence are fixed, for any sequence $S$ the total number $N_0$ ($N_0 = n_s + n_b$) of sites occupied by the solvent does not change by changing the conformation of the sequence. It follows that the partition function given by Eq. (3.1) can be rewritten as

$$Z(S,\Gamma) = \left(q_{ob}e^{-\beta E_{ob}} + q_{db}e^{-\beta E_{db}}\right)^{N_0} \times \left[\frac{\left(q_{os}e^{-\beta E_{os}} + q_{ds}e^{-\beta E_{ds}}\right)}{\left(q_{ob}e^{-\beta E_{ob}} + q_{db}e^{-\beta E_{db}}\right)}\right]^{n_s(\Gamma)} \quad (3.3)$$

The first term on the right-hand side being constant, it does not affect the value of the cost function $C(S,\Gamma)$. Therefore, the cost function only depends on the number of shell sites $n_s(\Gamma)$. In particular, it is easily shown that at temperatures at which the protein is stable, the smaller $n_s(\Gamma)$, the lower $C(S,\Gamma)$. Consequently, the native state of a protein has to be the conformation with the lowest number of shell sites $n_s(\Gamma)$ [53, 54].

Because of the lack of information (*a priori*) about general properties of the good sequences that could discriminate them from the rest, we tackle the problem by exact enumeration: both the sequence space and the conformation space are exhaustively investigated.

It is worth to mention that from the technical point of view, for this purpose we should in principle deal with all possible sequences of length $L$: considering the sequences as a binary code and assigning arbitrarily the value 0 to the polar monomers, respectively the value 1 to the hydrophobic amino-acids, the whole set of sequences is simply given by the binary numbers from $S_0 = 00\ldots00$ to $S_L = 11\ldots11$. Thus, for a given length $L$, the size of the sequence space is $2^L$ and all these sequences should be taken into account. However, this is not the case for our computations because of the properties of the generator of Self-Avoiding Walks. Indeed, our SAW algorithm, even if it does not take into account the conformations created by symmetries of rotation, it nevertheless keeps the ones obtained by the mirror symmetry. It follows that each good sequence would be taken into account twice if dealing with the whole set of sequences, since the reversed sequence matches with the mirror conformation of the first sequence. Therefore, instead of dealing with the whole sequence space, we first create a subset of sequences $\{S_L\}$: starting from the total number of sequences $2^L$, we delete each sequence that is the reverse of one already existing (obviously all palindromic sequences are present in the new set of sequences). Afterward, once the set is created we start with the computation of the values: given the subset $\{S_L\}$ of the sequences of length $L$, and the set $\{\Gamma_L\}$ of all the possible conformations, we look for those sequences $\{S'_L\}$ that have a unique state with a minimum $n_s(\Gamma)$ among all the possible conformations.

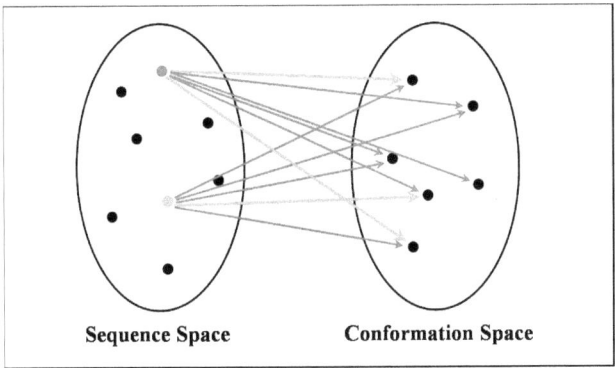

Figure 3.1: Schematic representation of the selection of *good* sequences. Each sequence $S$ is mounted on all possible conformations $\Gamma_i$ and the related cost function $C(S, \Gamma_i)$ is computed. For a given sequence, conformations with the lowest cost function (*i.e.* the *native states*) are represented by a green arrow. Sequences (represented by green dots) with a unique native state are retained and considered as *good* sequences, while those (red dots) with degenerate native states are discarded.

In addition, regarding the length $L$ of the investigated sequences, it is clear that the maximal length is limited by the computational power of the used computer and it is therefore important to correctly estimate the expected simulation time. On one hand, the size of the sequence space is obviously $2^L$ (neglecting the factor $\sim 2$ due to the elimination of the reverse sequences), since we deal with binary coded sequences. On the other hand, it can be shown that the number of Self-Avoiding Walks grows with a connective constant $\mu_{SAW} \sim 2.63$. For instance, for Self-Avoiding Walks made by $L = 18$ steps, there are 5 808 335 different conformations. Obviously, the computational time is directly proportional to the number of both sequences and conformations. It follows that the simulation time grows exponentially with the length $L$ of the investigated sequences and therefore exhaustive enumeration is limited to relatively short chains. Because of our restricted computational power we therefore limit our investigations by exact enumeration to sequences of length up to $L = 18$. However, as we will see, exploiting some intrinsic properties of the native states of the good sequences, allows to increase the length of the sequences to $L = 20$.

In order to achieve a large amount of data so as obtaining valuable statistical results, we deal with sequences of length $L = 10, \ldots, 18$. For each length $L$, all sequences $S_L$ are tested on all conformations $\Gamma_L$ and those sequences with a unique native state are retained: a new database is created containing, on one hand the set of good sequences $\{S'_L\}$ and, on the other hand, the corresponding set of native structures $\{\Gamma'_L\}$. The obtained values are shown in Tab. 3.1, where, for each length $L$, we indicate the sizes of the different sets of both sequences and structures. These results are discussed in the next two sections.

| Length | Total Sequences | "Good" Sequences | % | SAW | Native States | % |
|---|---|---|---|---|---|---|
| 10 | 528 | 42 | 7.9 | 2034 | 23 | 1.1 |
| 11 | 1056 | 95 | 9.0 | 5513 | 37 | 0.67 |
| 12 | 2080 | 200 | 9.6 | 15037 | 63 | 0.42 |
| 13 | 4160 | 379 | 9.1 | 40617 | 107 | 0.26 |
| 14 | 8256 | 676 | 8.1 | 110188 | 194 | 0.18 |
| 15 | 16512 | 1485 | 8.9 | 296806 | 364 | 0.12 |
| 16 | 32896 | 2431 | 7.4 | 802075 | 576 | 0.07 |
| 17 | 65582 | 5593 | 8.5 | 2155667 | 1068 | 0.05 |
| 18 | 131328 | 10050 | 7.6 | 5808335 | 1818 | 0.03 |

Table 3.1: Creation of the sets of *good* sequences. For each length $L$ (first column), starting from the whole set of sequences (second column), we retain those sequences (third column) possessing one unique native state (*i.e.* the so-called *good* sequences). The ratio between the two different sets is shown in the forth column and does not change significantly for different lengths $L$. In addition, the number of native states is indicated in the sixth column and compared with the total number of Self-Avoiding Walks (fifth column). It turns out that the ratio (seventh column) of native states decreases while increasing the length $L$. This result is in agreement with the concept of *fold*, the taxonomical classification of real proteins (see text).

## 3.2 Compactness of the native states

It is well known that folded globular proteins are characterized by both a compact three-dimensional structure and the presence of a hydrophobic core (the hydrophobic amino-acids of the chain try to hide their surface from the solvent). Since these properties are crucial for the stability of proteins, we begin with the investigation of the compactness of the native states obtained by our former design computation. For this purpose we define the *perimeter* $p_\Gamma$ (of a given conformation $\Gamma$) as the number of water sites that are in contact with at least one amino-acid and compute then the perimeter of each native state of our database.

As expected, we find that all sequences of the sets $\{S'_L\}$ (for $L = 10, \ldots, 18$) have native states $\{\Gamma'_L\}$ with their perimeter limited by some $p_{max,L}$. Moreover, for each length $L$, the difference between the perimeter $p_{min,L}$ of the most compact conformations and the largest perimeter $p_{max,L}$ is very small and does not exceed the value 3 (i.e. 3 more solvent sites in contact with the protein). These results (shown in Tab. 3.2) confirm that proteins designed with the HPW model possess very compact native states, results that are in complete agreement with the data of real proteins.

| Length | $p_{min}$ | $p_{max}$ | $\Delta(p_{min}, p_{max})$ |
|---|---|---|---|
| 10 | 12 | 13 | 1 |
| 11 | 12 | 14 | 2 |
| 12 | 12 | 15 | 3 |
| 13 | 13 | 15 | 2 |
| 14 | 14 | 15 | 1 |
| 15 | 14 | 17 | 3 |
| 16 | 14 | 17 | 3 |
| 17 | 14 | 17 | 3 |
| 18 | 15 | 18 | 3 |

Table 3.2: Distribution of the perimeter sizes. For each length $L$ (first column), the minimal possible perimeter $p_{min}$ is shown in the second column and the perimeter $p_{max}$ of the largest selected conformations is shown in the third column. It turns out that the difference $\Delta(p_{min}, p_{max})$ (last column) between the two values is very small: only compact conformations are chosen by the sequences as native states.

In addition, this empirical proof of the compactness of the native states allows to increase the length $L$ of the investigated sequences. Indeed, since structures are limited by some $p_{max,L}$, we can reduce the size of the conformation space by limiting the search of the possible native states among compact conformations. In particular, we apply this method for sequences of length $L = 20$: the $p_{min,20}$ being 16, tentatively, we investigate only those conformations with $p \leq 21$ (reducing drastically the size of the conformation space from 41 889 578 to 1 223 214).

With this approach we find a size $N_{S,20}$ of the set of good sequences $\{S'_{20}\}$ equal to 37 933 and the number of related native states $N_{\Gamma,20} = 5\,440$ (see Tab. 3.3).

| Length | Total Sequences | "Good" Sequences | % | SAW | Native States | % |
|---|---|---|---|---|---|---|
| 20 | 524800 | 37933 | 7.2 | 41889578 | 5440 | 0.01 |

Table 3.3: The sizes of the different sets for sequences of length $L = 20$.

In addition, we also find that the perimeters of the native states do not exceed $p_{max,20} = 20$: a result that support our initial guess of the maximal size of the needed perimeter. However, in order to obtain a further verification of our approach, we check if we have left apart some good proteins by extrapolating the number of sequences in $\{S'_{20}\}$ using the data for $L = 10, \ldots, 18$. Considering that the two related sets of walks on a lattice, i.e. the Hamiltonian Walks and the Self-Avoiding Walks, have both an exponential growth, we used an exponential fit for our data.

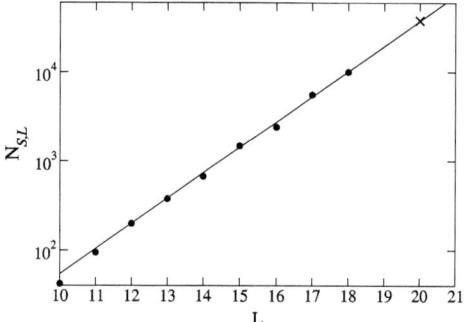

Figure 3.2: Number of sequences $N_{S,L}$ (the $y$ axis is logarithmic). The black dots represent the values found by exact enumeration for sequences of length $L = 10, \ldots, 18$. The line is the extrapolated curve, whereas the cross is the number of sequences in the set $\{S'_{20}\}$, created checking only the configurations with perimeter $p \leq 21$.

The extrapolated curve overshoots $N_{S,20}$ by 1%, which is compatible with the approximation of the curve for the lengths $L = 10, \ldots, 18$ (Fig. 3.2). This further check proofs that, even if in principle there could be other sequences in $\{S'_{20}\}$ that we did not find or degenerate competitors that we did not consider, still they should not represent a significant modification of the set.

## 3.3 Protein Folds and Designability

While the compactness of the native states seems to be a general property of all our proteins, we are furthermore interested in the behavior of the sizes of the different sets shown in Tab. 3.1 as a function of the length $L$. As expected, the number of *good* sequences increases with the length $L$ of the investigated sequences. However, the ratio between the number of good sequences and the total number of possible sequences does not change significantly by changing the length of the sequences. On the contrary, regarding the data of the structures, the ratio between the number of native states and the total number of possible conformations decreases exponentially while increasing the length $L$ (better represented in Fig. 3.3).

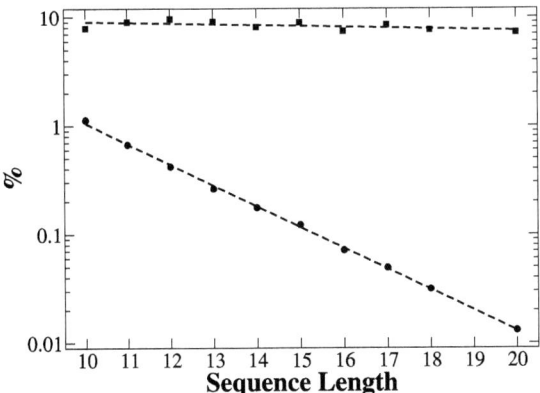

Figure 3.3: The ratio between the number of different conformations used as native states and the total number of SAW as a function of the length $L$ (black dots). For larger values of the length $L$ more and more *high designable* conformations are present so that less and less different conformations are needed as native states (see text). On the other hand, the percent of designable sequences with respect to the total number of sequences does not depend significantly on the length $L$ (black squares).

The presence of a small number of native states can be understood introducing the concept of *fold* or *family*: the taxonomical classification of real proteins. Indeed, proteins can be grouped following their similarity in their three-dimensional fold structures or in their amino-acid sequences. The standard hierarchical classification of proteins is in the form of *fold*, *superfamily* and *family* taxonomies. Each class is a subset of the one above. In the top level, proteins belonging to a common *fold* have their secondary structures occupying the same spatial arrangement. In the next level, proteins in the same *superfamily* are considered to share a similar protein function. In the third level, proteins are grouped into the same *family* if their amino-acid sequences are considered to be similar. It turns out that for real proteins the number of folds is extremely low compared to the number of known proteins: nature seems to prefer few particular structures. In particular, in the Protein Data Bank (PDB), while the number of classified proteins is around 20 000, the number of folds is only 701 [4]. Moreover, even if the number of determined structures increases rapidly, the number of "new folds" increases very slowly, suggesting that the total number of effectively different structures for real proteins is extremely low compared to the number of existing proteins (there are almost 120 000 determined protein sequences in the protein sequence database SWISS-PROT [62]).

A further important concept introduced in order to understand the preference for a few number of folds is the so-called *designability* of a three-dimensional structure: *high* designable structures are those chosen by a large number of proteins, while structures with a low parameter of designability are those adopted by a relatively small number of proteins. It has been shown that almost all possible three-dimensional structures have a low designability, while the number of conformations possessing high designability is rather small.

Both properties are also reproduced by the proteins of the HPW model. Indeed, on one hand, Fig. 3.3 shows that the ratio between the number of chosen conformation as native state is relatively small compared to the total number of Self-Avoiding Walks. Moreover, the ratio decreases with the length $L$ of the sequences. This happens because, for larger values of $L$, high designable conformations are present, reducing the ratio of needed structures, and creating larger fold families. On the other hand, also the inhomogeneous degree of designability is reproduced. Indeed, we checked the designability for the conformations of the native states of the sequences of length $L = 20$. We find that not all the compact structures have the same designability, with some of them more designable than others: many different sequences share the same native fold. Indeed, we find that 62% of all sequences in $\{S'_{20}\}$ have their native state on just the 17% of all designable conformations (those that are native states of 11 sequences or more; the highest designable structure attracts 147 sequences): most proteins find their native fold on a restricted number of structures. Moreover, as the protein length $L$ grows, the number of sequences per native structure increases: the ratio $\frac{N_{S,L}}{N_{\Gamma,L}}$ is proportional to $\mu_r{}^L$, with $\mu_r \simeq 1.12$ (see Fig. 3.4). This increase of the number of sequences per structure, is in agreement with the former result: for larger values of $L$ more and more high designable conformations become available and consequently chosen by a large number of sequences.

Finally, we also compute the growth of the number of designable conformations $\{\Gamma'_L\}$ as a function of the length $L$. It turns out that $\{\Gamma'_L\}$ grows with a connective constant $\mu_{des} \simeq 1.74$, value that is close to $\mu_{HW} \simeq 1.47$ typical of Hamiltonian Walks (SAWs have $\mu_{SAW} \simeq 2.63$). This result, together with the perimeter data, confirms that native states of our proteins are compact.

## 3.4 Cold and warm denaturation

After selection of the good sequences (creating the sets $\{S'_L\}$) with the design method, thermodynamic quantities for the sequences in $\{S'_L\}$ are easily computed. However, we first have to fix the parameter values (Fig. 3.5) of the MLG model considering some qualitative criteria.

Regarding the degeneracy parameters, as explained in the model description, the important criteria to be considered is their hierarchy: the values have to satisfy the inequalities $q_{ds} > q_{db} > q_{ob} > q_{os}$. First of all, the absolute multiplicity of the $q$ values being irrelevant (since contributing only with an additive constant to the free energy), we fix arbitrarily the degeneracy of the *ordered shell* sites $q_{os} = 1$. Furthermore, we choose large numbers for the degeneracy values $q_{ob}$, $q_{db}$ and $q_{ds}$. This choice is reasonable since every

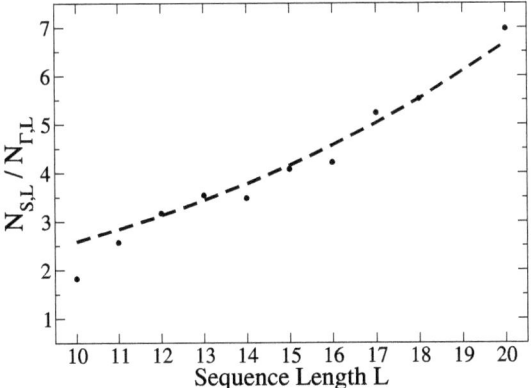

Figure 3.4: Ratio of sequences per conformations as a function of the length $L$. The points represent the computed values for the HPW sequences, while the dashed curve is the function $\mu_r{}^L$, with $\mu_r \simeq 1.12$. For larger values of $L$ high designable conformations are available and therefore chosen by a larger number of sequences.

site contains some water molecules, so that the total number of states per site will be the number of states of *one* single hydrogen bond to a power that is the *total* number of hydrogen bonds present in the site. Therefore, we fix the values as following: $q_{ob} = 2\,500$, $q_{db} = 7\,500$ and $q_{ds} = 10\,000$. It is important to stress that, even if these values are fixed rather arbitrary, a large amount of different values were tested. It turned out that the results shown in the next sections are robust in a range of various order of magnitude of the degeneracy parameters (which is reasonable since a large change in the $q$'s can correspond to even a slight change in the number of states for a single hydrogen-bond, due to the power raising).

As for the energy levels, once more, the parameters have to satisfy a set of inequalities, *i.e.* $E_{ds} > E_{db} > E_{ob} > E_{os}$. Since these parameters have no units, we fix the latter with a symmetry criterion, in order to decrease their degrees of freedom: $E_{db} = -E_{ob} = 1$ and $E_{ds} = -E_{os} = 2$. It is clear that a better determination of these values could come from molecular dynamics and structural studies. However, same as for the degeneracy parameters, we checked that the results do not change significantly by changing the energy parameters (with and without the symmetry constraint). In addition, this robustness was already pointed out in the literature [53, 54].

Once the parameters have been fixed, we begin with the computation of some thermodynamic quantities. Among them, of particular interest are the specific heat $C_v$ and

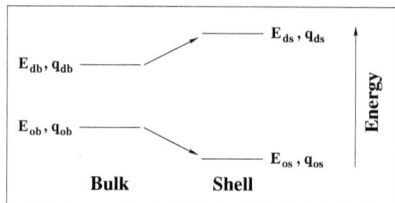

Figure 3.5: The MLG bimodal energy distributions for bulk and shell water molecules. The lower levels represent ordered groups of water molecules, the higher levels disordered ones.

the thermodynamical average of the number of shell sites $<n_s>$. Indeed, these two quantities computed as a function of the temperature $T$ give insight about the *stability* of a given protein. Therefore, we choose a sequence $S$ from the set $\{S'_L\}$ and by means of the partition function $Z(S) = \sum_\Gamma Z(S,\Gamma)$ where

$$Z(S,\Gamma) = \left(q_{ob}e^{-\beta E_{ob}} + q_{db}e^{-\beta E_{db}}\right)^{n_b(\Gamma)} \cdot \left(q_{os}e^{-\beta E_{os}} + q_{ds}e^{-\beta E_{ds}}\right)^{n_s(\Gamma)} \quad (3.4)$$

we compute both the specific heat

$$C_v(S) = \frac{d<E(S)>}{dT} \quad (3.5)$$

and the number of shell sites

$$<n_s(S)> = \frac{\sum_\Gamma n_s(S,\Gamma) Z(S,\Gamma)}{Z(S)} \quad (3.6)$$

as a function of the temperature $T$.

If the model reproduces the correct behavior of proteins, the shape of the specific heat $C_v$ should contain two peaks indicating two distinct transition phases. Indeed, it is known that, for real proteins, on one hand the free energy difference between unfolded states and the native state is positive below the *heat* (or *warm*) denaturation temperature $T_w$, ensuring the stability of the native state. While, on the other hand, this difference has a maximum around $20^oC$ for the "average" protein, and then decreases for temperatures $T < 20^oC$. Eventually, by lowing the temperature below a critical temperature $T_c$ it is even possible to see the *cold* denaturation of proteins [32]. It follows that these two denaturations, if present, are indicated by two peaks in the shape of the specific heat $C_v$.

In Fig. 3.6 we have chosen the particular sequence

$$PPPPPPHPPPHPPHPHHHHH$$

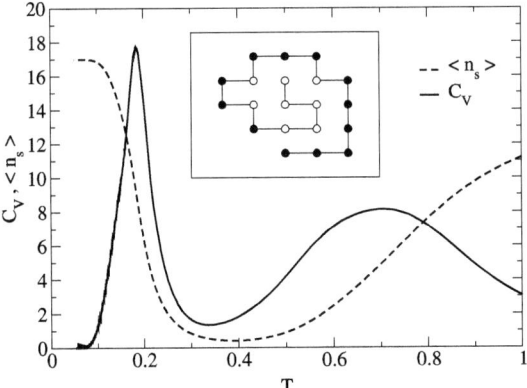

Figure 3.6: Specific heat $C_V$ (out of scale) and $<n_s>$ for the protein shown in the inset; $q_{ob} = 2\,500$, $q_{db} = 7\,500$, $q_{os} = 1$, $q_{ds} = 10\,000$, $E_{db} = -E_{ob} = 1$, $E_{ds} = -E_{os} = 2$. The two peaks in the specific heat $C_v$ indicate the presence of both *cold* and *warm* denaturations. Between them, the most probable conformation (shown in the inset) is the one with the minimal number of shell sites $n_s$ (*i.e.* the native state of the protein): the hydrophobic amino-acid (white spots) are hidden in the core of the protein and surrounded by the polar residues (black spots).

and the parameter values

$$\begin{aligned} q_{os} &= 1 \\ q_{ob} &= 2\,500 \\ q_{db} &= 7\,500 \\ q_{ds} &= 10\,000 \\ E_{db} &= -E_{ob} = 1 \\ E_{ds} &= -E_{os} = 2 \end{aligned}$$

(for the computations we have set the Boltzmann constant $k_B = 1$).

As expected, the specific heat $C_v$ and the thermal average $<n_s>$ as a function of $T$ point to two different denaturations, since the $C_v$ shape contains the two predicted peaks. Moreover, the shape of the thermal average $<n_s>$ shows that between the two peaks proteins have few contacts with water (they are in a compact state), and the most probable conformation is the one with the minimum $n_s(\Gamma)$ contacts (*i.e.* the native state as defined in the choice of the set of good sequences). In this conformation (see inset of Fig. 3.6), the hydrophobic amino-acids are hidden in the core of the protein. Above and below the two temperatures proteins swell and the number of water-protein contacts increases.

These results show that the HPW model captures within a single framework both *warm* and *cold* denaturations. It is important to stress that this behavior is not reproduced by most Hamiltonians used in protein folding, since their native state is the $T = 0$ ground state (GS) of the model (*e.g.* the HP model).

Finally, it is important to mention that, even if only the result for a particular sequence is shown, we tried various other sequences and lengths. As expected, it turned out that the results are always qualitatively the same, with slight changes in the peak height and width, and with some small $T_w$ and $T_c$ variations.

## 3.5 Statistical properties of the sequences

Design according to some model should be tested against as much of the known protein phenomenology as possible, at least qualitatively. In the former sections, we have shown that the HPW proteins already recover the correct thermodynamics, compactness and structure (segregation of a hydrophobic core) of real proteins. Some more information comes from sequence statistics.

Of particular interest for our model are however only those tests on which real proteins are described as a succession of hydrophobic and polar amino-acids. For this purpose, one has to be able to divide the 20 amino-acids into two groups; unfortunately, this division is not trivial. Indeed, while some amino-acids are clearly defined as hydrophobic (respectively polar), there are other residues for which the classification is more difficult. It follows that in the literature there is not a total agreement in the classification of the amino-acids. In particular, some scales prefer to assign an "hydrophobic value" to each amino-acid rather than using a strict binary code (*i.e.* the HP code). However, binary classifications exist: we use them in this section in order to look at two basic indicators, namely the *hydrophobic amino-acid concentration* and the so-called *Run Test*.

### 3.5.1 Hydrophobic amino-acid concentration

In our specific case, we use two commonly used hydrophobic scales: the Eisenberg and the Cornette scales [63, 64]. Starting from a sample (of size 2 111) of proteins chosen from the FSSP database (*Fold classification based on Structure- Structure alignment of Proteins* [65]), we assign to each amino-acid of a given sequence $S$ either the value H or the value P according to the classification of the two scales. Furthermore, once each sequence is rewritten in a binary code, we compute the concentration of hydrophobic amino-acid of each sequence and the distribution of the latter.

The two resulting curves (shown in Fig. 3.7) are well approximated by binomial distributions with mean values around $55-60\%$. Since according to these scales the number of hydrophobic amino-acid species is 11 and 12 respectively (55% and 60%), binomials peaked around these values are typical of *random* sequences. Consequently, the concentration of hydrophobic amino-acids of real proteins is indistinguishable from the one obtained by a set of binary sequences randomly created.

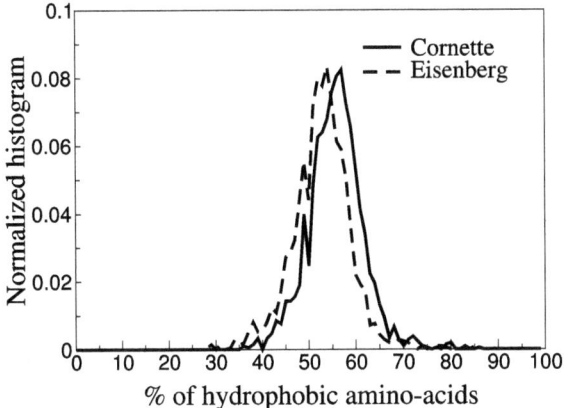

Figure 3.7: Distribution of the concentration of hydrophobic amino-acids for 2 111 proteins selected in the FSSP database. By means of both the Eisenberg and the Cornette hydrophobic scale, each sequence is rewritten in a binary code (H,P) and the distribution is plotted. Both curves are nicely approximated by a binomial curve peaked around 55%, respectively 60%. Since, according to the two different scales, the number of hydrophobic amino-acids is 11, respectively 12, the shapes of these curves are typical of those obtained by binary *random sequences*.

Since, under selective pressure, proteins have evolved toward very specific structures, at first glance it seems rather surprising to find some underlying randomness in their composition. However, one should not forget that there is an intrinsic randomness in the evolution process. For instance, mutations in the DNA sequences are completely random, but nevertheless crucial for evolution. It follows that some randomness is supposed to be preserved also in the present proteins.

On the other hand, it is clear that the randomness of the hydrophobic residue concentration does not give any insight about some special properties of proteins. However, it is important to stress that this is a part of the phenomenology of real proteins and as such it should also be recovered through models.

Therefore, we investigate both proteins of the HP and the HPW model in order to check if the latter possess the same distribution or, on the contrary, if one of the model may introduce an artificial bias on its protein composition. For this purpose, we compute, for the set of sequences of length $L = 16$, the number of hydrophobic residues for both models and we plot their distribution. The resulting curves are shown in Fig. 3.8.

Obviously, for a two-letter alphabet (P and H), random sequences possess a binomial distribution peaked around 50%. It turns out that, while this distribution is reproduced by the sequences of the HPW model, this is not the case for the sequence of the HP model. Indeed, Fig. 3.8 shows that the HP sequences have a distribution that, on one hand, is peaked around 60% and, on the other hand, not so well approximate by a binomial curve: the distribution is not in agreement with the one of real proteins. The HP model seems to prefer sequences with a larger number of hydrophobic amino-acids than nature does. On the other hand, the distribution of HPW sequences is nicely approximated by a binomial around 50%: the HPW model does not introduce any bias toward sequences with higher concentration of hydrophobic amino-acids than pure chance would do, in apparent similarity with real proteins.

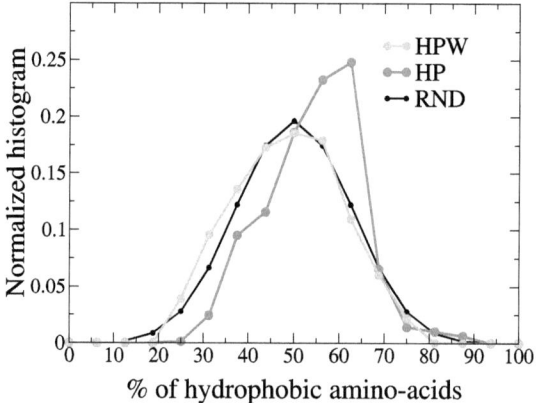

Figure 3.8: Distribution of the concentration of the hydrophobic amino-acids for sequences of length $L = 16$. While the HPW sequences behave similar to random sequences (and consequently to real proteins), the HP sequences have a preference for higher concentration of hydrophobic residues.

### 3.5.2 The Run Test

A further useful statistical test is the so-called *Run Test*, which has been introduced in the context of proteins by S.H. White *et al.* in 1994 [66]. Instead of testing the global concentration of hydrophobic amino-acids, the idea underlying this test is to check whether the *distribution* of hydrophobic residues along the chain of amino-acids follows some explicit rule. Same as for the former test, the investigated sequences have first to be reduced into binary strings of Hs (hydrophobic amino-acids) and Ps (polar amino-acids). Then, every series of consecutive Hs (or Ps) is counted as a *run*. As an example, the string HHHPP contains two runs (HHH and PP), and the string HPHPH contains five runs (each single letter counting as a run) (pictorially shown in Fig. 3.9).

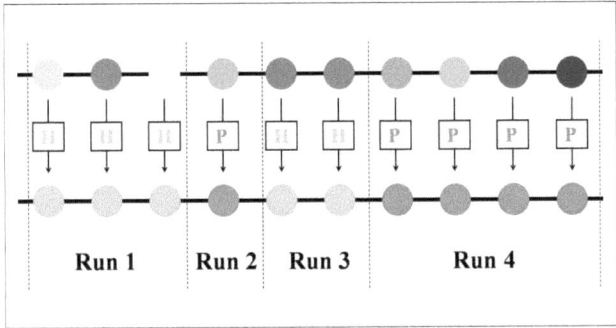

Figure 3.9: The Run Test: the original sequences are reduced into a binary code, assigning the value P to the polar amino-acids, respectively the value H to the hydrophobic residues. A *run* is defined as a series of consecutive Hs or Ps (single letters are also counted as runs).

Once the enumeration of runs is done, the Run Test consists in the analysis of the distribution of proteins according to the number of runs they contain. These computations have been done in former works for real proteins [66]. It turned out that, according to this test, real proteins are statistically indistinguishable from random sequences. Indeed, the distribution of the number of runs follows a binomial distribution, as expected from random sequences. It follows that, also for this further test, a set of real proteins is indistinguishable from one containing completely random binary sequences.

Same as for the results about the concentration of hydrophobic amino-acids, even if we do not want to dwell into the implications of this further result, it is nevertheless an intrinsic property of real proteins and should therefore be reproduced by any protein design model. For this reason we apply the Run Test to our sequences: Fig. 3.10 shows the distribution of the number of runs $n_r$ for proteins of length $L = 16$ designed using both the HP and the HPW model.

Since the largest possible number of runs for a sequence of length $L = 16$ is 16 (and, obviously, the smallest one being 1), the distribution for a set of random sequences follows a binomial curve peaked around 8.5. However, once more, sequences of the HP model failed the test. Indeed, Fig. 3.10 shows that the distribution of the number of runs $n_r$, even if peaked around 9, is nevertheless not symmetric, indicating that the HP model has a preference for sequences with many runs. On the other hand, the distribution of $n_r$ for the HPW proteins is again nicely approximate by the binomial curve that would be obtained by a set of random sequences: according to the Run Test, the HPW sequences behave similar as real proteins.

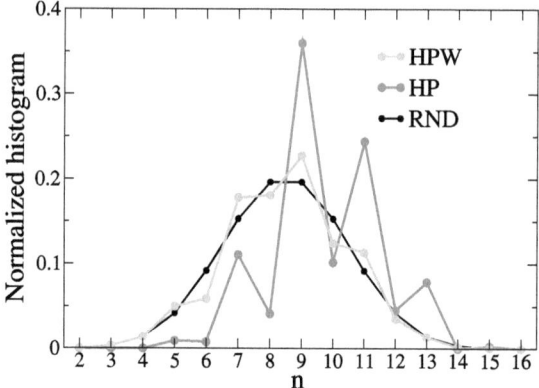

Figure 3.10: Distribution of the number of runs $n_r$ of the sequences of length $L = 16$. While the distribution of the HPW sequences is similar to the one of random sequences (and therefore to real proteins), the HP model has a preference for sequences with a large number of runs.

While these two tests give a further confirmation of some statistical similarities between the HPW sequences and real proteins, they furthermore allow an interpretation about the different behavior between the HP model and the HPW one. Regarding the HP sequences, the two tests indicate that sequences with a *high polar amino-acid density* have a high probability of having long P runs. Indeed, since in the HP model that we considered (*i.e.* the *standard* HP model) there are neither HP nor PP interactions, these long P runs would be free to fluctuate, giving rise to degenerate native states. It follows that these sequences have to be discarded during the design procedure (because of their degenerate native states) and, therefore, designable sequences have to possess both few Ps and many runs, in order to avoid fluctuations of the P runs [67]. The HPW model, instead, does not need such a selection since, at an effective level, the presence of the solvent introduces also HP interactions: the latter are able to avoid large fluctuations of any long P run and consequently no particular distribution nor particular concentration of hydrophobic amino-acids are needed.

Moreover, in order to check if our guess was correct, we also design proteins with a modified HP model. In this new model we fix the effective interaction values as follows:

$$\epsilon_{HH} = -2.3$$
$$\epsilon_{HP} = -1.0$$
$$\epsilon_{PP} = 0$$

With these new values, we first perform protein design for sequences of length $L = 16$ and, afterward, we apply to the new set of sequences both the concentration of hydrophobic amino-acid test and the Run Test (Fig. 3.11). It turns out that, indeed, the new proteins perform better than the ones of the standard HP model on both statistical tests, confirming the above interpretation: the introduction of HP interactions avoids fluctuations of long P runs.

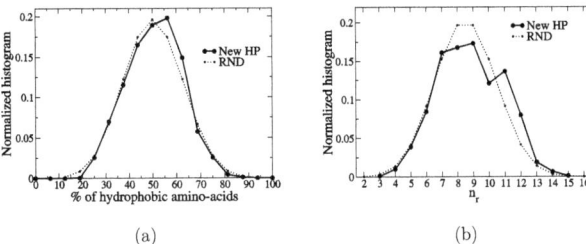

(a)  (b)

Figure 3.11: The modified HP model: statistical tests on the sequences of length $L = 16$. (a) Distribution of the concentration of the hydrophobic amino-acids. (b) Distribution of the number of runs $n_r$. Sequences of the new HP model perform better both statistical tests as the ones of the standard HP model, since possessing distributions closer to random sequences.

In summary, in this section, we showed that the HPW sequences, according to two statistical tests, namely the *concentration of hydrophobic amino-acids* and the *Run Test*, exhibit a similar behavior as real proteins. Indeed, for both tests it has been shown, that real proteins are indistinguishable from random sequences, property that has also been found by investigating the HPW sequences. On the other hand, it turned out that the HP sequences do not reproduce the same behavior. This difference is the consequence of the small number of used interactions, which rise to a high degeneracy of the native states for some particular kind of sequences. Finally, these results give a first insight about the richness of the effective interactions introduced by the solvent in the HPW model,

## 3.6 Overall Stability Criterion

In the first section of this chapter we introduced the concept of *protein design*, we applied it using the HPW model and we created new sets $\{S'_L\}$ of sequences, referred to as *good* sequences. As explained, proteins belonging to the $\{S'_L\}$ set have been selected to satisfy the criterion of *uniqueness* of the native state. Yet, it is not the only request that has to be imposed on proteins. Indeed, the native state has to be stable against other states: at equilibrium it should be the most favorable one. For Hamiltonians (such as in the HP model) where the native state is, at the same time, also the *ground state* of a protein, this condition is trivially satisfied for low temperatures, since below a critical temperature $T_c$ the ground state is, by definition, the one possessing the largest Boltzmann weight among all possible states. On the contrary, this is not the case for the HPW model, since the equilibrium state of a protein is situated between the two transition temperatures $T_c$ (cold denaturation) and $T_w$ (warm denaturation). It follows that the probability $p_s$ to occupy the native state between these two temperatures should be at least larger than 1/2 (1/2 being the value at the denaturation transitions). Not all the sequences in $\{S'_L\}$ satisfy this criterion, therefore a new set $\{S''_L\}$ has to be generated by selecting all those sequences $S \in \{S'_L\}$ such that $p_s > 0.5$ for $T \in [T_c, T_w]$.

Since the native state of a protein found by protein design may not be stable (crucial property for real proteins), we need to introduce a new definition of *native* state: the Boltzmann weight of the native state $Z(S, \Gamma_{nat})$, for a given sequence $S \in \{S'_L\}$, has to be larger than the sum over all partition functions of the excited states:

$$\left\{ Z(S, \Gamma_{nat}, T) > \sum_{\{\Gamma_{exc}\}} Z(S, \Gamma_{exc}, T) \right\} \bigg|_{T \in [T_c; T_w]} \quad (3.7)$$

This is a completely new criterion in protein design; indeed, as explained above, it is automatically satisfied by HP proteins. Once their native (ground) state is guaranteed to be unique, there will always be a temperature below which $p_s > 0.5$. In the HPW model, instead, the native state, even having the lowest cost function value, can result unfavorable compared to phase space regions of high cumulated probability. These regions are usually related with a high degeneracy of the first excited states. It follows that, instead of being folded, the protein may fluctuate among conformations possessing a higher cost function as the native state.

In order to investigate the properties of the sequences selected with this new criterion, we compute the new set $\{S''_{20}\}$ starting from the set of good sequences $\{S'_{20}\}$ (*i.e.* those of length $L = 20$). We find that, from the original 37 933 sequences present in $\{S'_{20}\}$, only 4 575 satisfy the *overall stability* criterion. However, the number of these sequences is large enough to apply the two statistical tests introduced in the former section.

It turns out that for the sequences of the new set $\{S''_{20}\}$ the concentration of hydrophobic amino-acids follows a binomial distribution peaked around 50%, meaning that, according to this test, also these sequences are indistinguishable from random sequences. However, the new sequences do not pass the Run Test. Indeed, as shown in Fig. 3.12, $\{S''_{20}\}$ proteins have a run distribution that is quite different from the distribution for random sequences. We nevertheless would like to preserve the statistical features of sequences as an indicator of the model validity. Hence, we need to revise the uniqueness and overall stability criteria to understand if it is possible to relax them, so to recover the good sequence statistics.

Protein folding is a dynamical process taking place at some temperature between the two transition temperatures $T_c$ and $T_w$. The *uniqueness* of the native state is usually invoked to ensure that the folding is not misled to a target structure different from the native one. Actually, chances are extremely good for correct folding even in the presence of (almost) degenerate competitors, if the basin of attraction (*i.e.* the *funnel* described in the introduction of the book) of the correct native state is much larger than that of the competitor (also referred to as *decoy*). We can easily envision the extreme case where

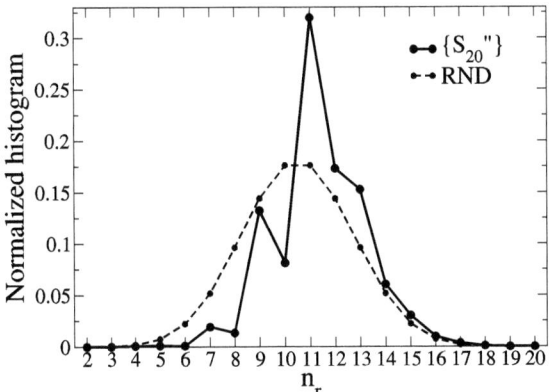

Figure 3.12: The Run Test for the new set of sequences $\{S_{20}'''\}$. The distribution does not follow a binomial curve, as it would be expected if the statistical properties of the new set would be similar to them of real proteins.

the native state has a very large basin of attraction, and there is a competitor that is like a golf hole in the free energy landscape. Clearly, this decoy will almost never be found by the dynamics, and this protein could be retained by natural selection, whereas, the same protein, using the strict criterion given by Eq. (3.7), would be discarded: although the Boltzmann weight of the competitor plus that of the other non-native states can be larger than the one of the native state, the competitor's weight should not be taken into account because, dynamically, it will almost never be found. Therefore, Eq. (3.7), with the exclusion of the competitor in the r.h.s., could be satisfied.

Unfortunately, our investigation is limited to thermodynamic quantities, consequently it does not give any information about the shape of the energy landscape. It follows that no reliable proof can be exhibited in order to support our guess about the "non-strict" application of the overall stability criterion (Eq. (3.7)). However, a detailed study of the structure of the conformation space, and of the *dynamical accessibility* of the native state and of its thermodynamic competitors could lead to a further definition of *native* state: sequences with a unique native state could be retained even if Eq. (3.7) is not satisfied. This happens in the case where the probability to dynamically fold in a thermodynamic competitor is negligible. Such new criterion would produce a new set $\{S_L'''\}$ of *good* sequences, which might keep the statistical properties of the set of sequences $\{S_L'\}$ (the one found using only the uniqueness criterion).

In summary, in this chapter we have approached the protein design problem using the HPW model and we have investigated some statistical and thermodynamic properties of the designed sequences. It turned out that the HPW proteins possess a very compact native state characterized by the presence of a hydrophobic core (typical behavior of globular proteins). Moreover, this state is only stable in a limited range of temperatures, since the model reproduces correctly the presence of two transition phases, *i.e.* the *cold* denaturation (at a temperature $T_c$) and the *warm* denaturation (at a temperature $T_w$). While the protein is compact between these two temperatures $T_c$ and $T_w$, above and below them, the protein swells and denaturates, in complete agreement with the behavior of real proteins.

Also the number of different structures chosen as native state was investigated. We found that sequences have a preference for some particular conformations (referred to as *high designable* structures), in agreement with the concept of protein *folds*.

Furthermore, using two different methods of sequence analysis, namely the concentration of hydrophobic amino-acids and the Run Test, we found that the HPW sequences do better reproduce statistical properties of real proteins as the sequences designed with the HP model.

Finally, using a new definition of *native* state, we created new sets of sequences and applied to them the two statistical tests. It turned out that also the design criteria themselves come under scrutiny, suggesting that a much more relevant role in protein folding should be given back to dynamics, and to a careful study of the structure of the phase space (the possible conformations and the way they are connected to each other by the dynamics) [70]. Unfortunately, these problems can not be tackled by our computations, since exact enumeration can not address dynamical issues, neither it can be used above 2D. It follows, that the exact definition of native state in the framework of the HPW model, is still an open issue.

# Chapter 4

# Chaotropic Cosolvent Effects

In the previous chapter we investigated properties of proteins designed with the HPW model and compared some of them with both those of real proteins and those of the HP proteins. In this chapter we introduce a modified HPW model in order to investigate the influence of *cosolvents* on the stability of proteins.

It has long been known that proteins can be unfolded in aqueous solution by high concentrations of certain reagents. In fact, the use of these particular substances in order to denaturate proteins is one of the primary methods of measuring the conformational stability of proteins. In particular, two reagents mostly used for experiments are *urea* and *guanidine hydrochloride* (GdnHCl). Indeed, the denaturation effect of urea has been known for 100 years [68], while the even greater effectiveness of GdnHCl was first reported by J. P. Greenstein in 1938 [69]. However, despite the widespread, almost centennial use of these denaturants, the mechanism underlying their action is still not well understood.

A common class of denaturants is represented by the so-called *chaotropic agents*. It is generally believed that these agents may decrease the stability of proteins either by binding directly to the amino-acid chain or by modifying the properties of the solvent [42, 43, 44, 45, 46, 47, 48, 49, 50, 51, 52]. While the direct binding of chaotropic molecules to proteins may result in the weakening of the hydrophobic interactions between hydrophobic amino-acids, it is believed that the insertion of a chaotropic cosolvent in the solution creates a perturbation in the hydrogen-bond network generated by the water molecules. Since in the previous chapter we have shown that the hydrophobic effect results from this partial ordering of water molecules, the perturbation induced by the cosolvent may consequently result in a weakening of the hydrophobic effect.

Therefore, in this chapter, in order to check the validity of this assumption, we introduce and we investigate a modified HPW model in which we assume that the main effect of chaotropic agents is to disorder water molecules [42, 71]. More precisely, the modifications in the model affect the degeneracies of the different possible states of the solvent, but we do not include a direct chaotrope-protein interaction. In this way, we do not consider hydrogen bonding of urea to the protein backbone, an effect that is most likely to occur [41, 46, 50, 51, 52].

## 4.1 The modified HPW model

In the original HPW model (henceforth referred to as *standard* HPW model), the water degrees of freedom are represented by Potts-like variables $\sigma$ that take on $q_{os} + q_{ds}$ or $q_{ob} + q_{db}$ values for shell or bulk sites, respectively (we recall that shell sites are defined as those in contact with hydrophobic amino-acids). The energy of a given amino acid sequence $S$ in conformation $\Gamma$ is

$$E(S,\Gamma) = \sum_{<i,H>} \left( E_{os}\tilde{\delta}_{i,os} + E_{ds}\tilde{\delta}_{i,ds} \right) + \sum_{(i,H)} \left( E_{ob}\tilde{\delta}_{i,ob} + E_{db}\tilde{\delta}_{i,db} \right) \quad (4.1)$$

where the first sum is over all shell sites and the second sum is over all bulk sites. (As usual, the subscripts refer to: $o$ = ordered states (water forms hydrogen bonds), $d$ = disordered states (broken hydrogen bonds)); $b$ = bulk, $s$ = shell. $\tilde{\delta}_{i,os} = 1$ if the Potts variable in site $i$ is in one of the $q_{os}$ ordered shell states, and $\tilde{\delta}_{i,os} = 0$ otherwise. Similar conventions for $\tilde{\delta}_{i,ds}$, $\tilde{\delta}_{i,ob}$ and $\tilde{\delta}_{i,db}$.

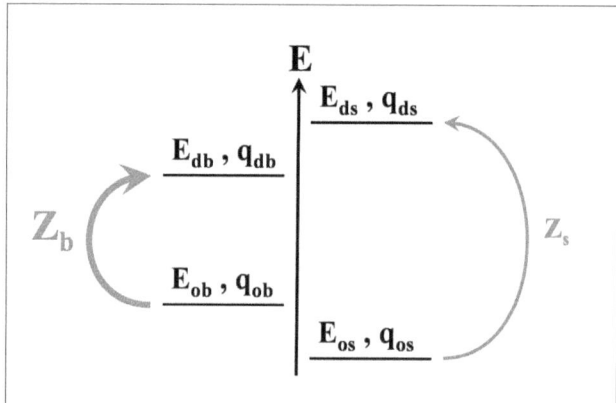

Figure 4.1: Schematic representation of the modified MLG model. By the addition of chaotrope agents in the solvent, ordered states change in disordered ones. In the model, the disrupting effect caused by chaotropes is regulated by the two parameters $z_b$ and $z_s$. Since the effect is much stronger in the bulk than in the shell, we have that $z_b \gg z_s$.

The partition function $Z(S,\Gamma)$ of a sequence $S$ in a conformation $\Gamma$ is

$$Z(S,\Gamma) = \left( q_{ob}e^{-\beta E_{ob}} + q_{db}e^{-\beta E_{db}} \right)^{n_b(\Gamma)} \cdot \left( q_{os}e^{-\beta E_{os}} + q_{ds}e^{-\beta E_{ds}} \right)^{n_s(\Gamma)} \quad (4.2)$$

where $n_s(\Gamma)$ is the number of shell water sites and $n_b(\Gamma)$ is the number of bulk water sites. Finally, the total partition function of a sequence S is given by

$$Z(S) = \sum_{\{\Gamma\}} Z(S,\Gamma) . \qquad (4.3)$$

In order to introduce chaotrope agents (also simply referred to as *chaotropes*) in the model we start from the assumption that their main effect is to disorder water molecules [42]: in presence of chaotropes the possible number of *ordered* states of the solute molecules decreases. More precisely, the introduction of a chaotrope cosolvent, since the total number $Q$ ($Q = q_{os} + q_{ds} = q_{ob} + q_{db}$) of states is fixed, imply a change of ordered states in disordered states. It follows that, in the modified model, a site in the bulk (shell) occupied by chaotropes will have $q_{ob} - \eta_b$ ($q_{os} - \eta_s$) ordered states and $q_{db} + \eta_b$ ($q_{ds} + \eta_s$) disordered states, where $\eta_b$ ($\eta_s$) is the number of bulk (shell) ordered states lost due to the disordering action of chaotropes.

Moreover, we also need to introduce a further parameter $c$ so as to express (and also to regulate) the *concentration* of the chaotrope agent. For this purpose, for a protein of given length $L$, we first fix the lattice size at $(M+L)$ sites, so that the total number (*i.e.* bulk + shell) of water sites is $M$. Afterward, we define the parameter $c$ as the chaotrope lattice concentration, so that the number of water sites occupied by chaotropes is simply $cM$. Moreover, we do not introduce any further distinction between bulk and shell sites, so that with this definition of the model, chaotropes can occupy both bulk and shell sites with the only constraint that the global concentration $c$ has to be constant. Therefore we can have $l$ shell sites and $(cM-l)$ bulk sites occupied by chaotropes and, consequently, for a conformation $\Gamma$ with $n_s(\Gamma)$ available shell sites, there will be

$$q(l,\Gamma) = \binom{n_s(\Gamma)}{l}\binom{M-n_s(\Gamma)}{cM-l} \qquad (4.4)$$

such distinguishable possibilities.

Starting from the original partition function of Eq. (4.2), we therefore introduce chaotropes in the model by writing the total partition function for a given sequence $S$ and a given concentration $c$ as

$$Z(S,c) = \sum_{\{\Gamma\}} Z(S,\Gamma,c) \qquad (4.5)$$

with

$$Z(S,\Gamma,c) = \sum_{l=0}^{n_s(\Gamma)} q(l,\Gamma) t_{bu}^{(cM-l)} t_{su}^{l} \times$$
$$t_b^{(M-n_s(\Gamma)-(cM-l))} t_s^{(n_s(\Gamma)-l)}$$

$$= t(c)\left(\frac{t_s}{t_b}\right)^{n_s(\Gamma)} \sum_{l=0}^{n_s(\Gamma)} q(l,\Gamma)\left(\frac{t_b t_{su}}{t_s t_{bu}}\right)^l \qquad (4.6)$$

where we introduced the simplified notations:

$$\begin{aligned}
t_b &= q_{ob}e^{-\beta E_{ob}} + q_{db}e^{-\beta E_{db}} \\
t_{bu} &= (q_{ob}-\eta_b)e^{-\beta E_{ob}} + (q_{db}+\eta_b)e^{-\beta E_{db}} \\
t_s &= q_{os}e^{-\beta E_{os}} + q_{ds}e^{-\beta E_{ds}} \\
t_{su} &= (q_{os}-\eta_s)e^{-\beta E_{os}} + (q_{ds}+\eta_s)e^{-\beta E_{ds}} \\
t(c) &= t_b{}^M \left(\frac{t_{bu}}{t_b}\right)^{cM}
\end{aligned}$$

for the different sites: $b$ = bulk, $s$ = shell, $bu$ = bulk with chaotropes, $su$ = shell with chaotropes.

With this new definition of partition function $Z(S,c)$ we are now able to compute various thermodynamic quantities of the protein solution as a function of the chaotropes concentration $c$. More specifically, since we are interested in the influence of chaotropes on the protein stability, we look at the free energy difference $\Delta G_{DN}$ between the native state and the denaturate states, value that we compute using:

$$\Delta G_{DN}(S,c) = -\frac{1}{\beta}\log\left(\frac{\sum_{\Gamma^S \neq \Gamma_n^S} Z(\Gamma^S,c)}{Z(\Gamma_n^S,c)}\right) \qquad (4.7)$$

where the $Z(\Gamma_n^S,c)$ is the partition function of the native state and the sum $\sum_{\Gamma^S \neq \Gamma_n^S} Z(\Gamma^S,c)$ is made over all other possible Self-Avoiding Walks ($\beta^{-1} = k_B T$).

Same as for the computations with the standard HPW model, the length $L$ of the investigated sequences is limited by our computational capabilities. We are therefore forced to restrict the investigations on sequences of length $L = 16$, so as allowing *exact* enumeration over all Self-Avoiding Walks (we recall that $N_{SAW,16} = 802\,075$). The sequences $S$ that we use are those of the set $\{S'_{16}\}$, *i.e.* those that have been designed so as to correspond to proteins with a unique native state in the absence of chaotropes. As for the other parameters of the MLG model, for our computations, we keep those that were used for the protein design in the former chapter, *i.e.*

$$\begin{aligned}
q_{os} &= 1 \\
q_{ob} &= 2\,500 \\
q_{db} &= 7\,500 \\
q_{ds} &= 10\,000
\end{aligned}$$

for the degeneracy parameters and

$$\begin{aligned}
E_{db} = -E_{ob} &= 1 \\
E_{ds} = -E_{os} &= 2
\end{aligned}$$

for the energies. However, we once more verified that the results are robust over a broad range of values (even many orders of magnitude for the parameters of the states number) as long as the inequalities among the energies and among the degeneracies are respected.

## 4.2 Protein stability

As introduced above, since the effect of chaotropic agents is the destabilization of the native state, we are interested in the free energy difference $\Delta G_{DN}$ between the denatured ($D$) and the native ($N$) states. Therefore, by means of Eq. (4.7), we compute $\Delta G_{DN}$ for different sequences of length $L = 16$. In particular, the following results are obtained using a sequence with a unique native state and such that no hydrophobic amino-acids are in contact with water in the native state, in the absence of chaotropes. The idea underlying the choice of a native state with a completely hided hydrophobic core is that the destabilization effect due to the introduction of the cosolvent may be amplified because of the larger difference of the number of shell sites between the native state and the denaturate states. Anyway, the results presented do not depend significantly on the particular sequence chosen.

All other parameters of the model being fixed (using the ones of the standard HPW model), we furthermore need to quantify the two disruption parameters $\eta_b$ and $\eta_s$. As for the water in the bulk, we assume that in order to denature the protein, chaotropic agents should reduce significantly the number of bulk ordered states of water. We quantify the disrupting effect of chaotropes on the order of water introducing the ratio $z_b = \eta_b/q_{bo}$ and use large values of $z_b$ to denature the protein. Also for the water molecules in the shell, we introduce the ratio $z_s = \eta_s/q_{so}$, value that measures the reduction of the shell ordered states of water due to the presence of chaotropic molecules. Since we assume that the disruption effect of ordered states in the bulk is stronger that the one in the shell, we also introduce a constraint on the two ratio, using values so that $z_b > z_s$.

In order to check the validity of the assumption that the destabilization effect is mainly due to the disruption of ordered states in the bulk, we start by computing the free energy difference $\Delta G_{DN}$, fixing the two parameters as follows:

$$\eta_b = 0.96$$
$$\eta_s = 0$$

That is, the cosolvent does not affect the ordering in the shell sites, but, on the contrary, disrupts strongly the ordered states in the bulk. Indeed, in this case, only 4% of the bulk ordered states is still available in the presence of chaotropes. With these parameter values, we therefore compute, for different concentration $c$ of the cosolvent the free energy difference $\Delta G_{DN}$ as a function of the temperature $T$. The resulting curves are shown in Figs. 4.2, 4.3: the destabilization due to the introduction of chaotropes in the solute is evident. On one hand, for $c = 0$ the obtained curve has the typical parabolic shape obtained with real proteins, denoting the presence of two temperature denaturations. On the other hand, Fig. 4.2 shows that at the temperature of maximal stability $T^*$ of the protein, for small concentration $c$ of the cosolvent, the native state is still stable (*i.e.* $\Delta G_{DN} > 0$). However, by increasing the concentration $c$, the protein becomes less and less stable. Finally, for values of $c$ larger than a critical concentration $c_D$, the protein denaturates (the free energy difference $\Delta G_{DN}$ becomes negative).

While Fig. 4.2 confirms that the destabilization effect due to chaotropes is mainly caused by the disruption of the ordered states in the bulk, we are furthermore interested

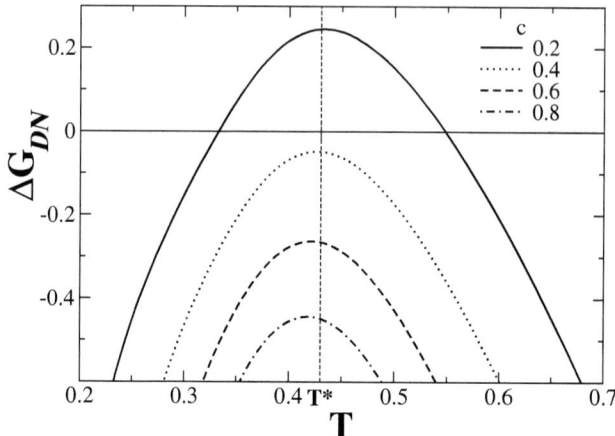

Figure 4.2: Free energy difference $\Delta G_{DN}$ between the denaturate state (D) and the native state (N) as a function of the temperature $T$ for different concentrations $c$. The ratio of ordered state lost in the bulk due to the presence of chaotropic molecules is $z_b = \eta_b/q_{bo} = 0.96$; the same ratio in the shell is $z_s = \eta_s/q_{so} = 0$. The denaturation is evident and it increases with $c$.

in the dependence of the behavior of the system on the shell ratio $z_s$. Therefore, we compute again $\Delta G_{DN}$ for fixed $z_b = 0.96$ and different values of $z_s$ (i.e. the ratio of disrupted ordered sites in the shell). It turns out that, for fixed $z_b$, as the concentration $c$ increases, denaturation occurs more easily for small values of $z_s$ (effect recognizable by comparing Fig. 4.2 ($z_s = 0$) with Fig. 4.4 ($z_s = 0.5$)). For these small values, the chaotropic agents do not influence much the order of the water molecules in the first hydration shell and the destabilization effect is strong. On the contrary, by rising the ratio $z_s$, this effect decreases and finally, once reached the extreme case of $z_b = z_s$, denaturation does not occur at all, for any concentration $c$ of chaotropic agents.

In order to understand these results, it is important to recall that around the temperature of maximal stability $T^*$, the free energy of the system is not dominated by a single term but it is rather a balance between an enthalpy gain/loss and an entropy gain/loss [53, 54]. Therefore, hydrophobic aggregation and protein collapse can be viewed as a result of the transfer of shell water molecules to the bulk, i.e. far from the contact with non-polar groups. Indeed, at physiological temperatures, the equilibrium state of shell sites is shifted toward disordered, high energy states, because they are many more than ordered ones. As a consequence, it is more convenient to turn these sites into bulk sites, that have a lower free energy because they have more low energy ordered states. On the other hand, since the disrupting effect induced by chaotropes diminishes the number of bulk ordered states, the effective gain of turning shell sites into bulk ones is reduced. It follows that the minimization of the number of shell sites is not longer favorable, as a consequence, the stability of the native state decreases drastically. However, as shown

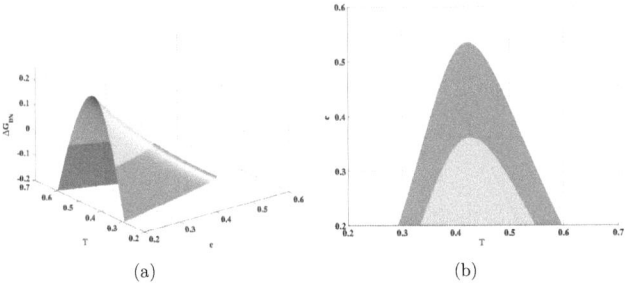

Figure 4.3: (a) Three-dimensional representation of the free energy difference $\Delta G_{DN}$ linearly interpolated from the data of Fig. 4.2. The zone of stability of the protein (green surface) shrinks drastically by increasing the concentration $c$ so that for concentrations $c$ larger than $\sim 0.35$ the protein becomes unfolded (red surface). (b) Two-dimensional projection on the axis $T$ and $c$ of the surface $\Delta G_{DN}$.

in Fig. 4.4, this effect is only present as long as chaotropes do not disrupt too much the shell ordered states.

Finally, we furthermore checked that large values of $z_b$ are needed in order to denaturate proteins. Indeed, as shown in Figs. 4.4, for small values of $z_b$, even if the effect is still present and the cosolute is able to destabilize the protein, the disruption of the ordered states is not strong enough to completely denaturate the protein. It follows, that two criteria have to be satisfied in order to obtain a complete denaturation. On one hand, the value of the ratio $z_b$ has to be large and, on the other hand, it has also to be considerably larger as the ratio for the shell $z_s$ ($z_b \gg z_s$).

In summary, in this first section, we have shown that the destabilization of proteins induced by the introduction of chaotropes in the solute, can be reproduced by assuming that the main effect of the cosolvent is the disruption of ordered states in the bulk sites. However, this effect is present as long as the disruption effect in the bulk sites is important and stronger as the the one in the shell sites.

## 4.3 The condensation of chaotropic molecules

While in the previous section we were interested in the destabilization of proteins induced by chaotropes, this section is devoted to the investigation of the behavior of the cosolute molecules. More precisely, we are interested in the comparison of the two cosolute concentrations in the shell and bulk sites. Indeed, it is worth to recall that with our definition of the model, we allow the $cM$ chaotropic sites to partition between bulk and shell sites as thermodynamically more convenient (see Eq. (4.6)). In order to obtain some information about this partition, we therefore compute the average concentration of chaotropic agents close to hydrophobic amino-acids $<c_s>$ and compare it with the global concentration $c$.

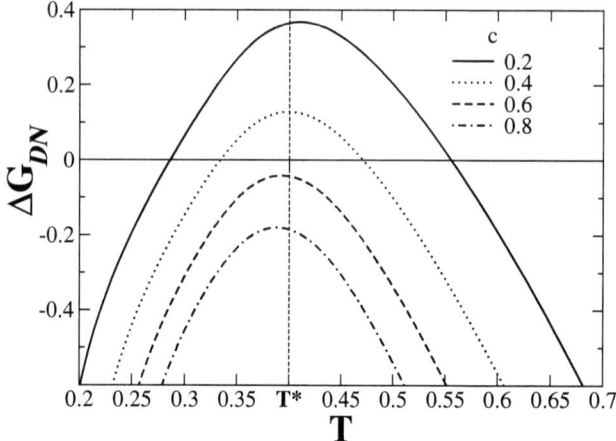

Figure 4.4: $\Delta G_{DN}$ as a function of $T$ for different concentrations $c$ and fixed parameters $z_b = 0.96$, $z_s = 0.5$. The energy gain obtained by forcing the chaotropic molecules in the shell sites decreases by increasing $z_s$: the denaturation effect of the chaotropes is weakened in comparison to the case $z_s = 0$ of Fig. 4.2.

Fig. 4.6 shows the behavior of the difference $< c_s > -c$ as a function of the global concentration $c$ and for different values of the ratio $z_b$. These values are computed at the temperature of maximal stability $T^* \simeq 0.4$ (see Fig. 4.4) and with a fixed disruption ratio for the shell sites $c_s = 0$. It turns out that around the temperature $T^*$ the average concentration of shell chaotropic molecules $< c_s >$ is always larger than $c$: chaotropes molecules are not uniformly distributed, but, on the contrary, they have a preference for shell sites. Moreover, the difference $< c_s > -c$ increases drastically with the disruption ratio $z_b$.

The fact that the concentration of chaotropes around hydrophobic monomers is higher than the bulk concentration can be explained looking at the behavior of the bulk water sites. Indeed, bulk liquid water tries to expel the chaotropes in order to not to be affected by its disordering action. This effect becomes more and more important as the disruption ratio $z_b$ becomes larger, as evidenced in Fig. 4.6. As a consequence we observe the condensation of chaotropic molecules around hydrophobic amino-acids, effect that is also observed in experiments and in simulations [52]. It is worth emphasizing that, within this model, such condensation is not due to any direct chaotrope-protein interaction, but, on the contrary, it is induced by the solvent behavior.

Moreover, we are furthermore interested to highlight that *preferential binding* is a general property of chaotropic agents. For this purpose we plot the average concentration $< c_s >$ as a function of the temperature $T$, for the extreme case $z_b = z_s$ (see Fig. 4.7). It turns out that even for this case, despite the fact that denaturation does not

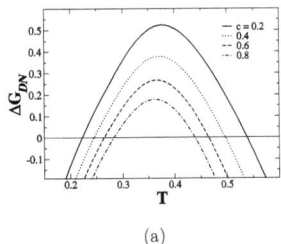 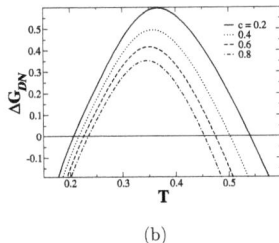

(a) (b)

Figure 4.5: $\Delta G_{DN}$ as a function of $T$ for different concentrations $c$ and fixed parameter $z_b = 0.8$ (a) Even for $z_s = 0$ the disruption effect is not strong enough to denaturate the protein. However, the destabilization effect due to the presence of chaotropes is evident. (b) For $z_s = 0.5$ the increase of the disruption effect in the shell sites decreases the destabilization effect, as expected from the former computations where destabilization occurs.

occur, at intermediate temperatures (when the protein is in its native state) the average concentration of chaotropes close to hydrophobic amino-acids $<c_s>$ is still higher than in the bulk. Only at low temperatures condensation does not occur, since, as shown in Fig. 4.7, the chaotropes molecules are uniformly distributed ($<c_s>=c$). The weakening of the condensation effect at low temperatures occurs because the limit of the term of the partition function (Eq. 4.6) for $T \to 0$ is:

$$\lim_{T \to 0} \left( \frac{t_b t_{su}}{t_s t_{bu}} \right) = 1 \qquad (4.8)$$

so that for $T \to 0$ all possible configurations become uniformly distributed and consequently:

$$\lim_{T \to 0} <c_s> = c \qquad (4.9)$$

Another interesting value, in order to understand the behavior of bulk sites is the value $c^*$, where the difference $<c_s>-c$ is maximal. Indeed, for $c = 0$ also $<c_s>=0$, and thus $<c_s>-c=0$; similarly, for $c=1$ also $<c_s>=1$. Therefore since $<c_s>-c$ is always non-negative, it should have a maximum for some value $c^*$ between 0 and 1 (see Fig. 4.7).

In Fig. 4.8 we plot the value $c^*$ as a function of the disruption ratio in the bulk $z_b$, at temperature of maximal stability $T^*$ and with the shell ratio $z_s = 0$. As expected, the value of $c^*$ decreases as $z_b$ increases. Indeed, since for large values of $z_b$ bulk sites are strongly affected by the disordering induced by chaotropes, even for small concentrations $c$ bulk sites try to expel most of the chaotropes. On the other hand, this effect is weakened decreasing the ratio $z_b$. As a consequence, the chaotrope concentration has to be large in order to affect significantly the bulk sites.

In summary, the results of this section show that the reduction of the number of ordered states for bulk water gives rise to an effective interaction between chaotropes and proteins. This effect is also observed for real proteins and it is therefore interesting that

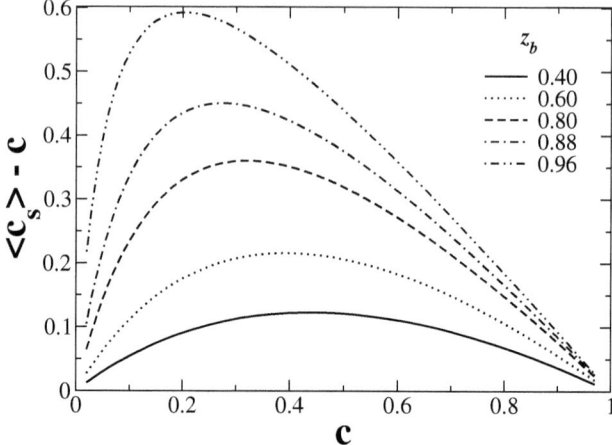

Figure 4.6: Behavior of $<c_s> - c$ as a function of $c$ for different values $z_b$ and fixed parameters $T^* = 0.4$, $z_s = 0$. In the temperature range of maximal stability, the average concentration $<c_s>$ is always larger than $c$ and the difference increases with $z_b$.

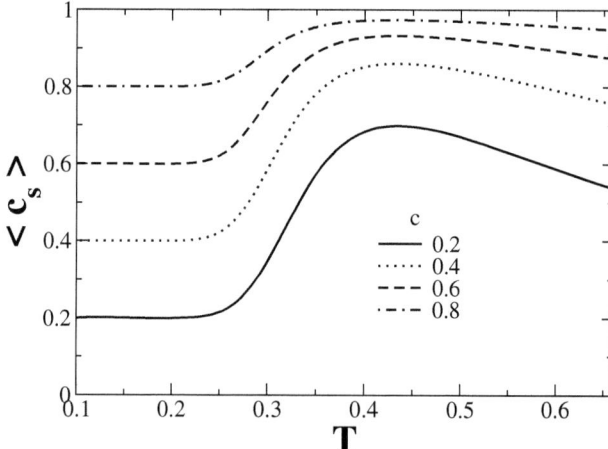

Figure 4.7: Concentration of the chaotropic agents in the shell $<c_s>$ as a function of $T$ and for different concentrations $c$ in the extreme case $z_s = z_b (= 0.96)$. For these values, the condensation does not occur at low temperatures, but it is anyway present at intermediate temperatures.

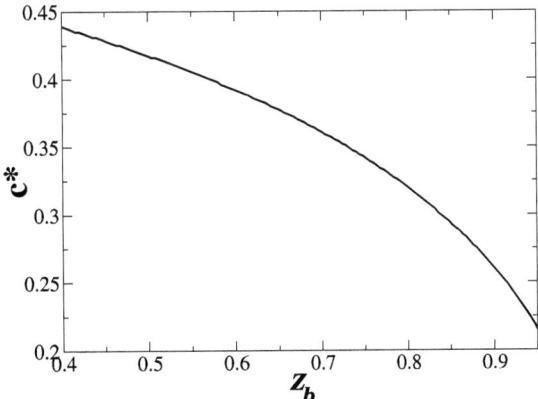

Figure 4.8: Concentration $c^*$, at which the difference $<c_s> - c$ has a maximum, as a function of $z_b$ and fixed parameters $T^* = 0.4$, $z_s = 0$. The $c^*$ value decreases by increasing $z_b$.

it can be reproduced by a simple model where no direct interactions between chaotropes molecules and proteins are introduced (even if those could further stabilize and enhance such effect).

## 4.4 Derivation of the $m$-value

Experimental observations highlighted the fact that the free energy of unfolding $\Delta G_{DN}$ ( i.e. the free energy difference between denaturate states and the native one) has, at a first approximation, a linear dependence on the concentration $c$ [72]:

$$\Delta G_{DN} = \Delta G_{DN}\big|_{c=0} - mc \qquad (4.10)$$

Consequently, the $m$-value is defined as [41]

$$m = -\frac{\partial \Delta G_{DN}}{\partial c} \qquad (4.11)$$

and, as observed above, for real proteins, it is not strongly dependent on $c$.

In order to verify if our model is as well consistent with this experimental result we therefore compute the behavior of the free energy difference $\Delta G_{DN}$ as a function of the global concentration $c$.

Fig. 4.9 shows $\Delta G_{DN}$ for different values of the ratio $z_b$. It turns out that, instead of obtaining a linear dependence on the concentration $c$ as expected from Eq. 4.10, we

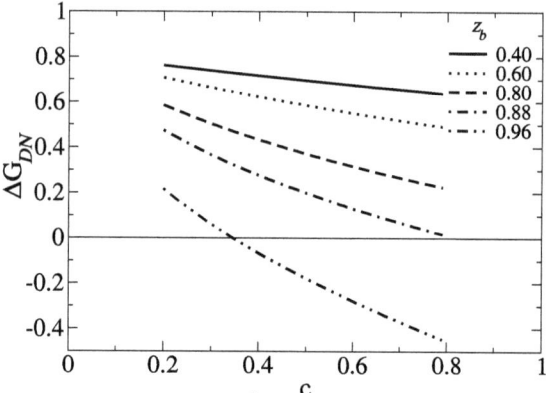

Figure 4.9: $\Delta G_{DN}$ as a function of $c$ for different values $z_b$ and fixed parameters $T^* = 0.4$, $z_s = 0$. An upward curvature is present for decreasing values of $c$.

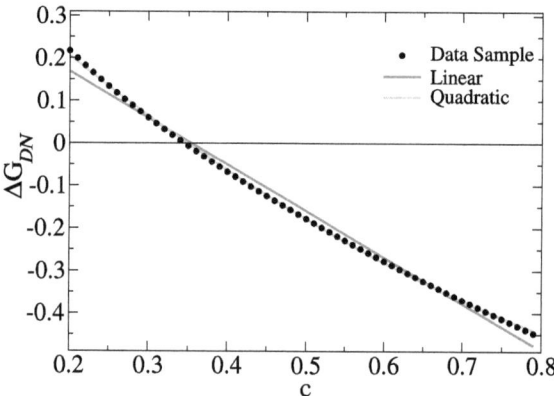

Figure 4.10: $\Delta G_{DN}$ as a function of $c$ for the worst case $z_b = 0.96$ and fixed parameters $T^* = 0.4$, $z_s = 0$. A linear fit gives a correlation coefficient $r_l = 0.995$ and a quadratic fit gives $r_l = 0.999$.

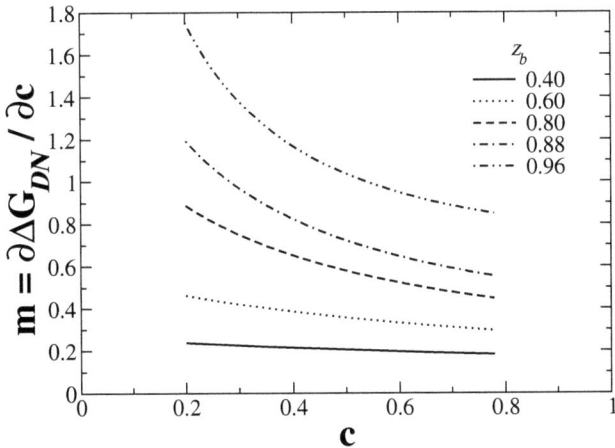

Figure 4.11: The $m$-value as a function of $c$ for different values of $z_b$ and fixed parameters $T^* = 0.4$, $z_s = 0$. The $m$-value is constant for small values of $z_b$. This is not the case for larger values of $z_b$, showing that the curvature of $\Delta G_{DN}$ increases with $z_b$.

observe an upward curvature in $\Delta G_{DN}$ for decreasing values of $c$. However, a similar upward curvature is also predicted by models that consider specific binding [45]. Moreover, even if small values of $c$ are difficult to observe experimentally since the protein is folded and extrapolation techniques should be used [41], a similar upward curvature was also observed in the case of urea denaturation of barnase [41]. Therefore, even if our results are not in complete agreement with Eq. (4.10), they are nevertheless similar to other experimental observations. In order to quantify this upward curvature, we fit the curve for the case $z_b = 0.96$, i.e the one with the smallest curvature. The curve is better fitted with a quadratic equation, indeed a linear fit of $\Delta G_{DN}$ gives a correlation coefficient $r_l = 0.995$, while a quadratic fit gives $r_q = 0.999$ (see Fig. 4.10). From the data computed for Fig. 4.10 we afterward also compute the $m$-value using Eq. (4.11). The obtained values are plotted in Fig. 4.11. While for small values of the ratio $z_b$ the $m$-value does not depend from the concentration $c$, as predicted from experimental observations, the behavior of the $m$-value changes for larger values of $z_b$. Indeed, Fig. 4.11 shows that the curvature of the $m$-value increases with $z_b$ and its dependence on the concentration $c$ becomes important for large values of $z_b$, i.e. for those values where the disrupting effect is important.

The $m$-value is however also dependent on other values. In particular, it is known that the $m$-value is related to $\Delta ASA$: the change in the accessible surface area ($ASA$) upon unfolding [73]. More precisely, in [73] 45 proteins were investigated, proteins that have been chosen because possessing $m$-values obtained from denaturation experiments and for which the structure of the native state is known. For these proteins, the $\Delta ASA$ were determined by the differences in solvent accessibility of the native state (computed from the crystal structure) and the unfolded form (as modeled by an extended polypep-

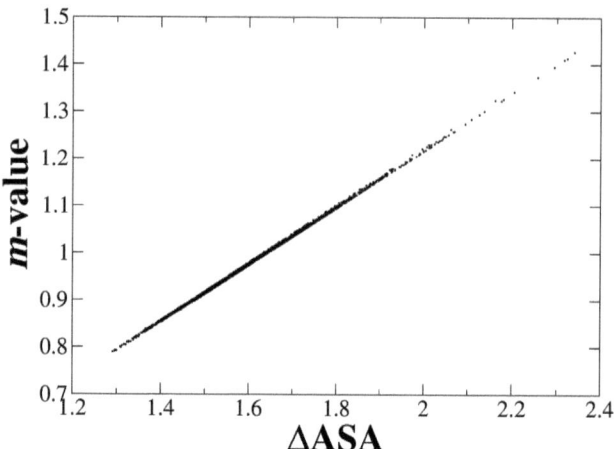

Figure 4.12: The $m$-value as a function of $\Delta ASA$ and fixed parameters $T^* = 0.4$, $z_s = 0$, $z_b = 0.96$ and $c = 0.6$. For each sequence we compute the $\Delta ASA$ (the change in the surface exposed upon unfolding) and the $m$-value. The linear fit for this data gives a correlation coefficient $r_l = 0.9997$, in agreement with experimental results.

tide chain). The data were computed for two different denaturants, namely *urea* and *guanidine hydrochloride* and the $m$-values were plotted as a function of $\Delta ASA$. It turned out that for both cases there is a significant linear correlation, with an $R$ value of 0.87 for Gdn HCl and 0.84 for urea.

We are therefore interested to check if our model is able to reproduce the same behavior. To compute $\Delta ASA$ we assume that our system has a two-states behavior, *i.e.* the protein can be either folded or unfolded. Consequently, we define as *folded state* the native state of the sequence, whereas the *unfolded state* is represented by all excited states. Therefore, the partition function of the unfolded state contains all possible conformations except the native one. Moreover, for each possible conformation $\Gamma$ of a sequence $S$ we compute the $ASA(\Gamma^S)$ using the definition of perimeter introduced in Chapt. 3: the total number of solvent sites in contact with the protein. With these definitions, for a given sequence $S$, we compute $\Delta ASA(S)$ using:

$$\Delta ASA(S) = \sum_{\Gamma^S \neq \Gamma^S_n} \frac{ASA(\Gamma^S) \cdot Z(\Gamma^S)}{\sum_{\Gamma^S \neq \Gamma^S_n} Z(\Gamma^S)} - ASA(\Gamma^S_n) \qquad (4.12)$$

so that we can compute, for each sequence $S_i$, its $m$-value using Eq. (4.11) and the $\Delta ASA(S_i)$ using Eq. (4.12).

Since the $m$-values are directly related to the denaturation properties of cosolvents, for our computations we fix the disrupting ratio in the bulk $z_b = 0.96$, the one in the shell $z_s = 0$ and the total concentration of the cosolvent $c = 0.6$, so as emphasizing the denaturation effects (as shown in Fig. 4.2). Fig. 4.12 shows the result computed at the temperature of maximal stability $T^* = 0.4$ for the whole set of sequences $\{S'_{16}\}$ (*i.e.* the sequences of length $L = 16$ with a unique state). Remarkably, the data of our model show a very strong linear dependence, since the linear fit for our data gives a correlation coefficient $r_l = 0.9997$. This result is therefore in complete agreement with the one found for real proteins.

## 4.5 The $(T, c)$ phase diagram

While in the former sections we mainly worked at fixed temperature $T$, in this section we are interested in the general behavior of the concentration of cosolvent $<c_s>$ in the shell sites. More precisely, we investigate the dependence on the two parameters $T$ and $c$ of the condensation of chaotropic molecules around hydrophobic groups, studying the phase diagram of the concentration of chaotropic agents of the shell sites $<c_s>$ and the related *phase transitions*. Indeed, if we consider system conformations where the chaotropic molecules are uniformly distributed both in the bulk and in the shell sites (*i.e.* $<c_s> \simeq c$) as a *gas phase*, we can define the *liquid phase* as the one represented by states where condensation is present (*i.e.* $<c_s> \gg c$). Consequently, the observed phase transitions (if present) will be of liquid-gas type.

First of all, in order to quantify the relation $<c_s> \simeq c$, we fix $\frac{<c_s>-c}{c} \leq 0.01$ as upper limit for the gas phase. Consequently the liquid phase is represented by all those conformations where $\frac{<c_s>-c}{c} > 0.01$.

Fig. 4.13 shows the computed phase diagram of a particular sequence $S$ for different values of $z_b$ and $z_s = 0$. As expected, phase transitions are observed. Indeed, Fig. 4.13 reveals that for large values of the temperature $T$, the chaotropic agent is in the gas phase and the chaotrope molecules are uniformly distributed. On the other hand, in the low temperature region, condensation is present and the concentration of cosolvent $<c_s>$ in the shell sites is larger than the global concentration $c$. While the presence of two states is a general property of the system and, in particular, it does not depend from the disruption ratio $z_b$, the phase transition temperature $T_c$ between liquid and gas is not constant. In fact, Fig. 4.13 shows that the temperature $T_c$ increases with $z_b$. This behavior can be explained in terms of entropy: the expulsion of chaotropic molecules from a single bulk site implies a free energy gain that increases with $z_b$. Indeed, it is possible to show that this free energy gain is proportional to $z_b$. On the other hand, by increasing the temperature $T$, the entropic advantage of being in a gas phase suddenly overcomes the sum of the single free energy gains. Finally, since the single free energy gain increases with $z_b$, the larger is $z_b$, the higher is the temperature needed for the entropic term $(T\Delta S)$ to overcome this free energy gain.

However, it is worth to stress that the phase transition temperature $T_c$ is always much larger than the heat denaturation temperature $T_{heat}$. Therefore, in the temperature range of protein stability, condensation is always present and the system is in the liquid phase. It follows that, since $T_c \gg T_{heat}$ for every concentration $c$ and every disruption ratio $z_b$, the *preferential binding* is a strong and relevant feature of our model and, moreover, it

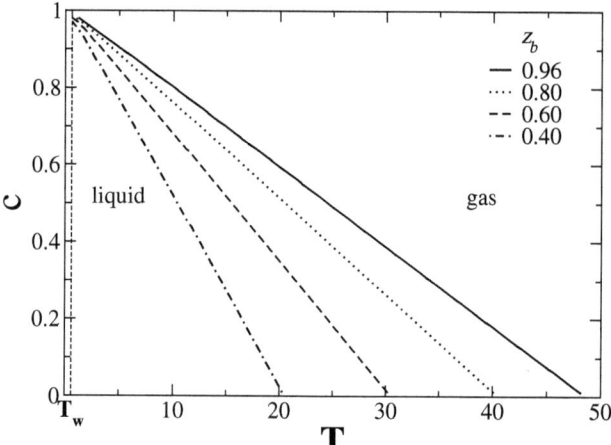

Figure 4.13: $(T, c)$ phase diagram for the concentration of chaotropic molecules on the surface of the protein. At low temperatures $T$ and low concentrations $c$, chaotropic molecules are concentrated on the protein surface (liquid-like phase). At high $T$ and high $c$, chaotropic molecules are uniformly distributed in the solution (gas-like phase). We show the phase diagram for different $z_b$ values and for $z_s = 0$. The transition temperature $T_c$ between the liquid and the gas phase increases with $z_b$.

does not depend significantly on the details of the latter.

In conclusion, in this chapter we have shown that the alteration of the network of hydrogen-bonds between water molecules due to the presence of chaotropic agents induces destabilization of proteins, and leads, in the extreme case, to complete denaturation, as observed in experiments. Moreover, our results evidence that this effect is important for large values of the disruption ratio $z_b$ in the bulk, but it is weakened by increasing the disruption ratio $z_s$ in the shell sites. In addition, we find that the reduction of the number of bulk ordered states due to the presence of chaotropes originates an effective interaction between the chaotropes and the protein. As a result, the chaotropic molecules accumulate near the surface of the protein. The effect is due to the fact that there is a free energy advantage when the chaotropic molecules are close to the surface of the protein since the reduction of ordered states of the surface is less than that of the bulk. As for the phase diagram of the system, we studied the average concentration of cosolvent $<c_s>$ present in the shell sites as a function of the chaotrope concentration $c$ and the temperature $T$. It turned out that, at low temperatures, the effective interaction between proteins and chaotropes induces the condensation of chaotropic molecules near the protein surface, while in the high temperature region chaotropic molecules are uniformly distributed in the aqueous solution.

Finally, we mention that it is likely that, beyond the simple model that we use, chaotropic and protein molecules may also interact directly [46, 51, 52, 74]. However, we have shown that chaotropic agents may destabilize protein native states even in the absence of direct interactions.

# Chapter 5
# Solvent effects on effective interactions

As mentioned in the introduction, several approaches have been attempted to the protein folding problem, since *all-atoms Molecular Dynamics* has been successful only for small proteins [11]. Indeed, it is important to recall that one of the major problem on protein folding is the fact that the process takes place in the presence of an aqueous solution. The latter has therefore to be taken into account in order to reproduce correctly the folding process. A common method to overcome the complexity of the solvent is to describe the solvent in a *implicit* form. In particular, one technique is the determination of effective amino-acid interactions for coarse-grained models of proteins. Indeed, if it would be possible to reproduce the solvent effects by means of amino-acid interactions, the complexity of the system would be drastically reduced. The purpose of this chapter is therefore to investigate, by means of the standard HPW model, the validity of this approach.

## 5.1  Effective interactions on real proteins

The aim of this approach is to obtain an interaction value for each possible amino-acid couple, which is afterward utilized to generate a pairwise Hamiltonian that should be able to stabilize the native state of any given protein. Starting from known data, for example taken from the Protein Data Bank, the idea is rather simple: the value of each couple interaction is determined examining the relative positions of the two amino-acids forming the considered couple. Fig. 5.1 is a very simple cartoon that shows the idea of the method. In this figure, the residues Valine and Leucine are always close each other. Therefore, an attractive energy will be assigned to this couple. On the other hand, the chain avoids couples of Lysine residues: a repulsive interaction will be assigned to this other couple. Finally, some couples of amino-acids do not interact (or weakly interact), since the distribution of their relative positions is very broad. No interaction energy will be therefore assigned to these kinds of couples. If the amount of data is large enough, this process can be performed for all 210 possible couples, creating a $20 \times 20$ matrix used to generate a pairwise Hamiltonian. Obviously, this is only a hand-waving picture. For the real case several techniques have been developed in order to determine these interactions. Among them, of particular interest is the *Perceptron* algorithm [16], where a simple

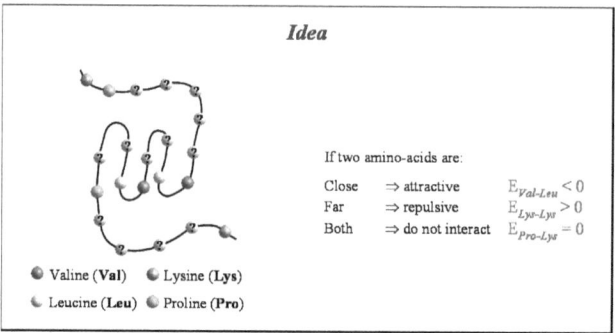

Figure 5.1: Schematic representation of the idea underlying the effective interaction method. Starting from known native states (left), an interaction value is assigned to each possible couple of residues (right), depending on their relative positions in the structures. In this simple example the couple Valine-Leucine would possess an attractive interaction, since being always close. The couple Lysine-Lysine a repulsive interaction and the couple Prolyne-Lysine no interaction at all, since their relative positions are not constant.

neural-network tries to learn contact interactions from a given training set of sequences and structures [17].

To apply a perceptron method to protein folding (or protein stability), one has first to take three main choices:
(a) the representation of the protein structure
(b) the cost function
(c) a set of alternative structures among which the native state is selected.
As for the representation of structures, the easiest technique for an effective interaction method is to deal with a *contact map* representation of protein structures. In general, a contact map is a square matrix $C$, whose elements are defined as:

$$C_{ij} = \begin{cases} 1 & \text{if residues } i \text{ and } j \text{ are in contact.} \\ 0 & \text{otherwise.} \end{cases} \quad (5.1)$$

where the definition "to be in contact" depends on the used description. For instance, the simplest form to define a contact is to consider two amino-acids to be in contact if their $C_\alpha$ atoms are closer than a fixed threshold $R_c$ (referred to as $C_\alpha$ *definition*). However, also more detailed definitions have been employed, such as, for example, the so-called *all-atoms definition*: two amino-acids are in contact if *any* two atoms (belonging to two different residues) are closer than a fixed threshold $R_c$ (note that this definition usually does not take into account hydrogen atoms). Once the definition of contact is chosen, the next step is to choose a suitable cost function. In the framework of effective interactions, the most general definition of cost function (assumed to be an approximation of the *true free energy*), can be written as:

$$E(S, C, \vec{\epsilon}) = \sum_{i<j}^{N} C_{ij}(S)\epsilon_{p_i p_j} \qquad (5.2)$$

where the $C_{ij}(S)$ are the elements of the matrix $C$ defined in Eq. 5.1, while the $\epsilon_{p_i p_j}$ are the effective interaction values (*i.e.* $\epsilon_{p_i p_j}$ represents the energy gain obtained if a residue of kind $p_i$ and one of kind $p_j$ are in contact). Finally, alternative structures (referred to as *decoys*) are needed, since the perceptron algorithm (as we will see in more details in Sub-Sec. 5.2.1) learns the interaction values by comparison of the energies of the native state and the ones of the decoys. Is is important to mention that the creation of decoys is a critical task, since the efficacy of the perceptron is strongly dependent on the "quality" of the chosen decoys. In order to create decoys, the most used method is the so-called *threading* method: given a chain of $N$ residues, decoys are generated using structures constituted by exactly $N$ residues of known protein structures composed by $N'(> N)$ amino-acids. However, it is important to stress that decoys generated by threading are not guaranteed to be physical! Indeed, because of the different sizes of the residues, in general, a given amino-acid sequence cannot be squeezed onto a structure belonging to another sequence. Finally, once the native state, the energy function and the decoys are fixed, the perceptron algorithm can be applied: a training set of proteins with their related native structures and decoys are used to learn effective interaction values for which the energy of the native state is the lowest among all decoys.

The perceptron method has been widely used with a broad range of different definitions of contact maps and different kinds of decoys. However, even if this technique has turned out to be one of the most promising methods, the computed effective interactions are able to stabilize only small sets of proteins [18]. While dealing with this method, it turns out that the perceptron is able to generate effective interaction values that are normally able to single out the native states of the proteins used for the training process, but are not efficient for proteins not belonging to the original set. Actually, in some cases, the values are also able to stabilize some proteins not included into the training set. However, the number of these additional proteins is of the order of units. Moreover, the method is successful only if the size of the training set is not too large. Typically, this size does not exceed the order of ten units [18]. In general, different training sets give rise to different effective interaction values. The latter are effective for proteins of their own sets but not applicable to other proteins (this result is pictorially shown in Fig. 5.2). It is important to recall that the leading idea of the method was to generate a set of *global* effective interactions, *i.e.* a set of values that would be able to stabilize any given protein. However, as demonstrated in all these works, this attempt failed.

In addition, while effective interactions are able to stabilize the native structure (*i.e.* the interactions forbid the protein to unfold), it has been demonstrated that a simple pairwise energy function may not *fold* even a single protein such as crambin, which is a very short protein since composed by only 46 residues [17]. Indeed, in this work it has been shown that with a $C_\alpha$ atom contact map and considering only pairwise contacts, it was not possible to find a set of effective interactions such that the native state possesses the lowest energy among all other possible conformations.

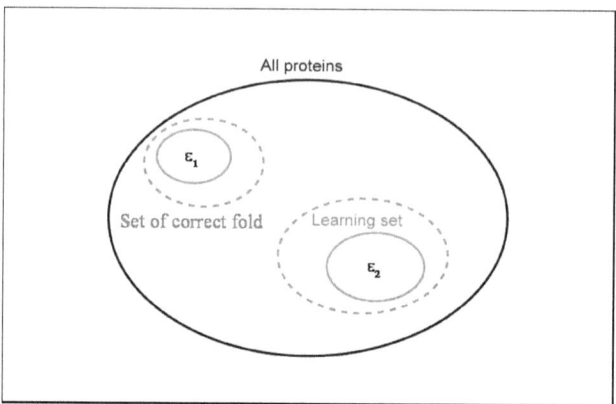

Figure 5.2: Schematic representation of the results obtained by the perceptron method. Different learning sets (blue ovals) give rise to different sets of effective interaction values $\{\epsilon_i\}$. However, the latter are only able to stabilize the proteins of the related sets (or, in the best case, few more proteins (red ovals)). It follows that it is not possible to generate a set of *global* interactions able to stabilize the whole set of proteins (black oval).

All these results clearly reveal that there is still a lack of understanding in the folding process: the coarse-grained models, used for the interaction extraction, are not able to capture the essential features of the forces present in the process. However, it is believed that these results may be improved if additional energy terms, such as hydrophobic, hydrogen bonds, or multibody interactions, are included.

## 5.2 Effective interactions on lattice models

As explained in the previous section, the effective amino-acid interactions used to define energy functions are not accurate enough to fold proteins, and it was therefore argued that the addition of further energy terms may lead to the correct folding process. In this chapter, in order to check the validity of this assumption we inspect, by means of a perceptron algorithm, if the hydrophobic effects induced by the solvent may be expressed by some kind of effective amino-acid interactions. More precisely, we apply the perceptron to proteins of the HPW model and try to generate a new Hamiltonian composed by amino-acid interactions for which the native state of a protein is still the one with the lowest energy.

We recall that the simplest model taking into account the hydrophobicity of non-polar residues (by means of effective interactions) is the HP model [55, 56], where the hydrophobic effect is captured by an attractive interaction between hydrophobic amino-acids (*i.e.* with an *implicit* solvent description). The energy of a given sequence $S$ mounted on a conformation $\Gamma$ is:

$$H(S,\Gamma) = \sum_{i<j} \epsilon^{nn}_{p_i p_j} \Delta_{nn}(\vec{r}_i, \vec{r}_j), \tag{5.3}$$

where $\epsilon^{nn}_{p_i p_j}$ are the interaction potentials (*e.g.* $\epsilon^{nn}_{HH} = -1$ and $\epsilon^{nn}_{HP} = \epsilon^{nn}_{PH} = \epsilon^{nn}_{PP} = 0$ for the *standard* HP model) while $\Delta_{nn}(\vec{r}_i, \vec{r}_j)$ is the usual nearest-neighbors contact matrix:

$$\Delta_{nn}(\vec{r}_i, \vec{r}_j) = \begin{cases} 1 & \text{if } i,j \text{ are nearest-neighbors} \\ 0 & \text{otherwise.} \end{cases} \tag{5.4}$$

The HPW model borrows the main properties of the HP one, with the difference that it has an *explicit* solvent description. If follows that, if the solvent effects may be described by effective interactions, also for proteins designed with the HPW model it should be possible to find a set of effective interactions. The latter should lead to a pairwise contact Hamiltonian (similar to the one of the HP model) that is able to identify the original native states (*i.e.* those obtained with the original HPW model). In order to check the validity of this hypothesis we therefore deal with the same techniques used for real proteins and we apply a perceptron algorithm to HPW proteins.

### 5.2.1 The perceptron

The original perceptron algorithm was invented by F.Rosenblatt on 1962 and it is the simplest neural network, since composed by only one neuron [16]. This algorithm is a binary classifier that assigns any possible input to one of two classes $A$ and $B$. By means of a learning program the correct assignment of the inputs to the correct class improves, *i.e.* as more the neural network learns, as more it will become reliable.

An input of a perceptron is in form of a vector $\vec{x}$ of a $n$-dimensional space: each component of this vector quantify, in mathematical terms, qualities of the object that has to be classified. The output is a scalar quantity $y(\vec{x})$.

The main point of this algorithm is that the output $y(\vec{x})$ also depends from a specific parameter of the perceptron called *weight* and expressed as a $n$-dimensional vector $\vec{w}$. More precisely, the output $y(\vec{x})$, corresponding to the input $\vec{x}$, depends on $h(\vec{x})$, the *scalar product* between $\vec{x}$ and $\vec{w}$:

$$h(\vec{x}) = \vec{x} \cdot \vec{w} = \sum_{i=1}^{n} x_i w_i \tag{5.5}$$

and is given by

$$y(\vec{x}) = \begin{cases} 1 & \text{if } h(\vec{x}) \geq 0 \\ 0 & \text{otherwise.} \end{cases} \tag{5.6}$$

Therefore, if we assign arbitrary the value 1 to the class $A$ and the value 0 to the class $B$, any object $\vec{x}$ will be classified in one of the two classes according to its output value $y(\vec{x})$. The goal of the perceptron is to find a weight vector $\vec{w}^*$ such as, for any given input $\vec{x}$, the output $y(\vec{x})$ is correct. In geometrical terms, the aim of the perceptron algorithm is to

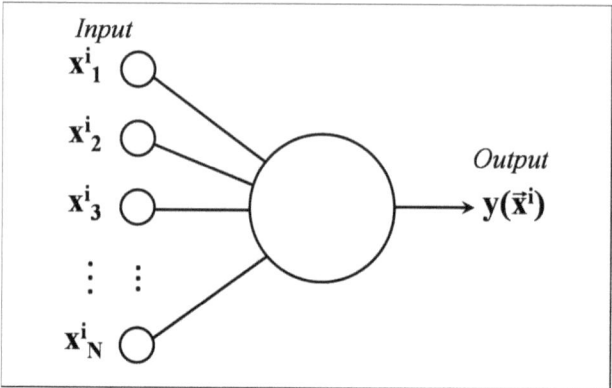

Figure 5.3: Schematic representation of the perceptron. Each input is described by a vector $\vec{x}^i$ quantifying, in mathematical terms, qualities of the represented object. The output $y(\vec{x}^i)$ is a scalar quantity, usually converted in a binary form. Any object $\vec{x}^i$ is classified into two classes depending on the value of the output $y(\vec{x}^i)$.

find a $(n-1)$-dimensional hyperplan (defined by means of the weight vector $\vec{w}$, the latter being its normal vector) that divides the whole set of input vectors $\vec{x}_i$ into two subspaces (i.e. the two classes $A$ and $B$), so that each subspace contains only vectors $\vec{x}_i$ that give rise to the same output. This geometrical point of view is pictorially shown in Fig. 5.4 for the 2-dimensional case.

In order to generate this hyperplan, the perceptron needs a so-called *training set*, i.e. a set of $n$-dimensional vectors $\vec{x}_i$ whose expected correct outputs $y^{exp}(\vec{x}_i)$ are known. Depending on the output of the training vectors $\vec{x}_i$ the algorithm is able to adjust the weight vector $\vec{w}$ and to make it converge toward the expected vector $\vec{w}^*$ (i.e. the vector for which each output is correct). The algorithm is an iterative procedure. Starting from a trial weight vector $\vec{w}_0$, the training vectors $\vec{x}_i$ are sequentially presented and, at each step, the perceptron produces an output $y(\vec{x}_i)$ according to Eq. (5.6) and it is compared to the expected output $y^{exp}(\vec{x}_i)$. In case of agreement the perceptron moves to the next training vector $\vec{x}_{i+1}$, otherwise the weight vector $\vec{w}$ is modified in order to correct the mistake. Many different rules are employed to update the weight vector $\vec{w}$. Here we just introduce the simplest one: at each step $i$, the weight vector $\vec{w}_i$ is updated following the rule:

$$\vec{w}_{i+1} = \vec{w}_i + \eta \Delta y(\vec{x}_i) \vec{x}_k / |(\vec{w}_i + \eta \Delta y(\vec{x}_i) \vec{x}_k)| \tag{5.7}$$

where $\vec{x}_k$ is the input vector chosen at time $i$, $y(\vec{x}_k)$ is the output defined by Eq. (5.6), $\Delta y(\vec{x}_i) = y^{exp}(\vec{x}_i) - y(\vec{x}_i)$ and $\eta$ is an arbitrary correction parameter. The effects of this rule are pictorially shown in Fig. 5.5(a) for the 2-dimensional case. In this picture, at step $i$ the object $x_9$ is not in the correct subspace. The weight vector $\vec{w}_i$ is therefore corrected by a quantity $\eta \vec{x}_9$ so that at step $i+1$ the hyperplan defined by the weight vector $\vec{w}_{i+1}$

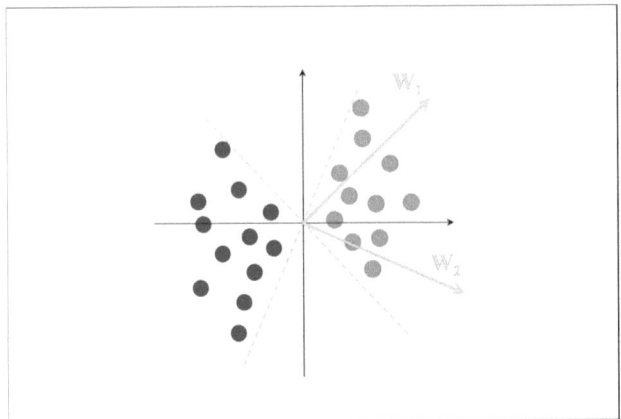

Figure 5.4: Schematic representation of the idea underlying the perceptron in the case $n = 2$. The aim of the perceptron is to find an hyperplan (whose normal is the weight vector $\vec{w}$) so that all objects are correctly classified. In this case, all objects of the same color has to belong to the same subspace: any weight vector $\vec{w}$ that lies between the two vectors $\vec{w}_1$ and $\vec{w}_2$ is a solution.

correctly classifies all objects $x_i$. In addition, this schematic representation is also useful to point out that such hyperplan may also not exist. Indeed, in the example of Fig. 5.5(b) no hyperplan exists able to correctly classify all objects $x_i$, since the two objects $x_3$ and $x_9$ belong to different subspaces for any weight vector $\vec{w}$.

It follows that one of the two following cases occurs: either the given problem is such as shown in Fig. 5.5(a), *i.e.* a solution exists (this case is referred to as *learnable problem*), either no hyperplan exists able to divide correctly the input space: the problem is called *unlearnable*. However, the crucial and interesting point of the method is that, in the case of existence of a solution, a convergence theorem guarantees that the perceptron will find it in a *finite* number of training steps, regardless the initial trial weight vector $\vec{w}_0$ and the correction factor $\eta$. It is worth to mention that in general, different vectors $\vec{w}_0$ lead to different points of the space of solutions. For instance, in Fig. 5.4 any vector that lies between the vectors $\vec{w}_1$ and $\vec{w}_2$ is a solution.

## 5.2.2 The perceptron on lattice proteins

In order to apply the perceptron method for the computation of effective interactions, we need to rewrite the problem in a appropriate form. As introduced above, the general form of a pairwise contact Hamiltonian, for a sequence $S$ mounted on a conformation $\Gamma$,

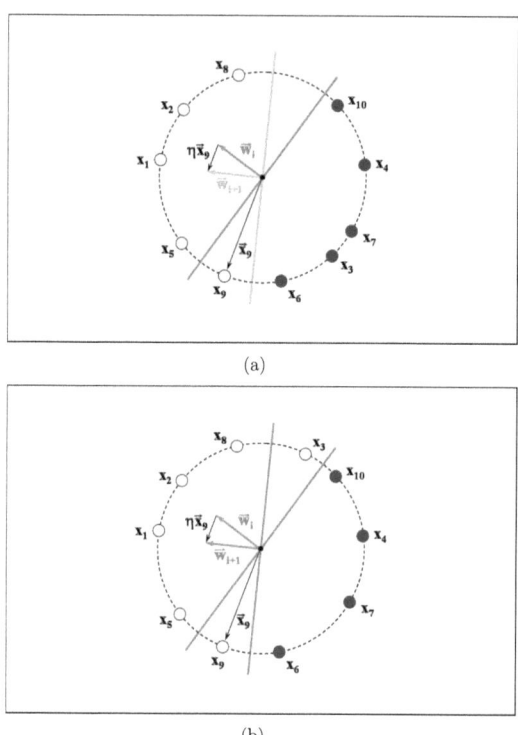

(a)

(b)

Figure 5.5: Schematic representation ($n = 2$) of the update of the weight vector $\vec{w}$, in case of a wrong output. (a) Learnable problem. At step $i$, since the object $x_9$ is not in the correct subspace, the weight vector $\vec{w}_i$ is updated by a quantity $\eta \vec{x}_9$. The hyperplan defined by the new weight vector $\vec{w}_{i+1}$ classifies correctly all objects. Indeed, the iteration of this update moves the object $x_9$ more and more in the middle of the correct subspace. (b) Unlearnable problem. For this example no such hyperplan exists. Since the two objects $x_3$ and $x_9$ belong to different subspaces for any weight vector $\vec{w}$. The problem is said to be not *linearly separable*.

is given by:

$$H(S,\Gamma) = \sum_{i<j} \epsilon_{p_i p_j} \Delta(\vec{r}_i, \vec{r}_j), \tag{5.8}$$

where the $\epsilon_{p_i p_j}$'s are the interaction values and $\Delta(\vec{r}_i, \vec{r}_j)$ is a contact matrix. In the case of simple nearest-neighbors contacts (*e.g.* as in the HP model), this Hamiltonian can be rewritten in another form introducing the notation

$$\epsilon_1 = \epsilon_{HH} \qquad \epsilon_2 = \epsilon_{HP} = \epsilon_{PH} \qquad \epsilon_3 = \epsilon_{PP} \tag{5.9}$$

Indeed, with this notation we can rewrite Eq. (5.3), for a sequence $S$ on a configuration $\Gamma$, as:

$$H(S,\Gamma) = \sum_{i=1}^{N} \epsilon_i C_i(S,\Gamma), \tag{5.10}$$

where $N = 3$ and $C_i(S,\Gamma)$ is the number of nearest-neighbors contacts of type $i$ ($i$ is mapped so as to match with the indexes given by Eq. (5.9)). It is important to stress that, while the following mathematical developments are done for the simple nearest-neighbors case, this form of Hamiltonian is nevertheless completely general and does not depend on the number of used interactions. In particular, it does not change introducing further interactions such as, for example, *next-nearest neighbors* interactions. If the interaction values are correct, the native configuration $\Gamma$ of each sequence $S$ has to possess the lowest Hamiltonian value among all other possible conformations $\Gamma_d \neq \Gamma$, *i.e.*

$$H(S,\Gamma_d) - H(S,\Gamma) > 0 \qquad \forall \Gamma_d \neq \Gamma \tag{5.11}$$

Therefore, using the Hamiltonian given by Eq. (5.10), the inequalities of Eq. (5.11) can be rewritten as:

$$\sum_{i=1}^{N} \epsilon_i C_i(S,\Gamma_d) - \sum_{i=1}^{N} \epsilon_i C_i(S,\Gamma) > 0$$

$$\sum_{i=1}^{N} \epsilon_i \{C_i(S,\Gamma_d) - C_i(S,\Gamma)\} > 0$$

$$\sum_{i=1}^{N} \epsilon_i C_i^\Delta(S,\Gamma_d,\Gamma) = \vec{\epsilon} \cdot \overrightarrow{C^\Delta}(S,\Gamma_d,\Gamma) > 0$$

$$\forall \Gamma_d \neq \Gamma \tag{5.12}$$

where we introduced the quantity

$$C_i^\Delta(S,\Gamma_d,\Gamma) = C_i(S,\Gamma_d) - C_i(S,\Gamma) \tag{5.13}$$

*i.e.* the difference of contact number of kind $i$ between configuration $\Gamma$ and $\Gamma_d$. Finally, because for a set of $M$ sequences the interaction potentials should not depend on the chosen sequence, Eq. (5.12) becomes:

$$\vec{\epsilon} \cdot \overrightarrow{C^\Delta}(S,\Gamma_d(S),\Gamma(S))) > 0 \qquad \begin{array}{l} \forall \quad S = 1,...,M \\ \forall \quad \Gamma_d(S) \neq \Gamma(S) \end{array} \tag{5.14}$$

In these inequalities the vector $\overrightarrow{C^\Delta}(S, \Gamma_d(S), \Gamma(S))$ only depends from the topology (i.e. the amino-acid positions) of the configurations $\Gamma_d$ and $\Gamma$. It follows that for a lattice model this vector is easily computed if both the native state $\Gamma$ and the decoy $\Gamma_d$ are known. In addition, the vector $\vec{\epsilon}$ contains the amino-acid interaction values: each component represents a different interaction (for instance, if we use the same notation as in Eq. (5.9), the first component will be the nearest-neighbors HH interaction).

The form of Eq. (5.14) reveals that the problem of finding a set of effective interactions such that the native state possesses the lowest energy can be converted into the problem of finding a vector $\vec{\epsilon}$ such that all scalar products on the left-hand side of Eq. (5.14) are positive. As we saw above, this task can be accomplished by a perceptron algorithm. Indeed, we can identify the vector $\vec{\epsilon}$ of Eq. (5.14) with the weight vector $\vec{w}$ of Eq. (5.5) and impose that all input vectors $\overrightarrow{C^\Delta}(S, \Gamma_d(S), \Gamma(S))$ have to belong to the same subspace. The latter constraint is obtained introducing a simpler update rule for the weight vector $\vec{w}$:

$$\vec{w}_{i+1} = \begin{cases} (\vec{w}_i + \eta \vec{x}_k)/|(\vec{w}_i + \eta \vec{x}_k)| & \text{if } y(\vec{x}_k) = 1 \\ \vec{w}_i & \text{otherwise.} \end{cases} \qquad (5.15)$$

From this it follows that if the perceptron is able to find a vector $\vec{\epsilon}_c$ such that all vectors $\overrightarrow{C^\Delta}(S, \Gamma_d(S), \Gamma(S))$ are in the same subspace, all inequalities of Eq. 5.14 are trivially satisfied, finally meaning that a set of interactions was found that is able to recognize the native states $\Gamma$ of the proteins. On the other hand, the absence of a solution implies that the parametrization of the Hamiltonian is wrong.

### 5.2.3 The perceptron on the HP model

In order to check the validity of the perceptron algorithm we apply the method to the standard HP model in the two-dimensional case, investigating sequences of length $L = 16$. Indeed, since HP proteins are designed according to an effective interaction Hamiltonian, if the method is correct the perceptron should be able to reproduce the same interaction values used for the original design.

As introduced above, in order to apply the method, for each sequence $S_i$, in addition to the native state $\Gamma_n(S_i)$, we also need a set of decoys $\Gamma(S_i)$. While, as opposite to real proteins, the creation of decoys for a lattice model is a simple task, since, a priori, each accessible conformation is known, some care has nevertheless to be taken while choosing them. To be stable, the native state of a protein has to possess the lowest energy among all possible conformations. In particular, reminding the funnel picture of the energy landscape, competition may mainly take place between the native state and similar conformations with slight higher energy. It follows that it is reasonable to select decoys of the first excited states (we recall that for each sequence $S$ the energy of any conformation $\Gamma(S)$ is, a priori, known). However, since the excited states are high degenerate, we randomly choose, for each sequence in $\{S_i\}_{16}$ (i.e. the set containing the sequences with a unique native state) only 15 decoys from the first three excited states (more precisely, we choose 5 decoys for each excited state). Once the decoys have been selected, the vectors $\overrightarrow{C^\Delta}$ of Eq. (5.14) are easily computed using the contact matrix of

Eq. (5.4) and starting from a random weight vector $\vec{\epsilon}_0$ we iterate the learning process. At each step $i$, all scalar products of Eq. (5.14) are computed and the vector $\overrightarrow{C_L^\Delta}$ with the lowest scalar product is identified. The weight vector $\vec{\epsilon}_i$ is then updated following:

$$\vec{\epsilon}_{i+1} = \vec{\epsilon}_i + \eta \overrightarrow{C_L^\Delta} \tag{5.16}$$

This particular rule allows to increase the convergence speed [75]. Indeed, the vector $\overrightarrow{C_L^\Delta}$ is the one that, in some sense, is the "worse" classified. Updating the weight vector according to Eq. (5.16) will therefore involve the largest possible modification of the hyperplan in order to place all vectors in the same subspace. This efficacy is easily understood by means of Fig. 5.5: vectors placed around the center of the wrong subspace are those with the lowest scalar product, since the weight vector is pointed in the opposite direction. Therefore, Eq. (5.16) is the update that generates the largest rotation of the hyperplan. Finally, by iterating the process, if a solution exists, i.e. a vector $\vec{\epsilon}_c$ exists satisfying all inequalities of Eq. (5.14), the value $\vec{\epsilon}_i \cdot \overrightarrow{C_L^\Delta}$ becomes positive in a finite number of steps [16]; otherwise the algorithm runs forever.

As expected, in the case of the standard HP model, the perceptron recovers the correct interaction values, i.e. $\epsilon_{HH}^{perc} \cong -1$, $\epsilon_{HP} \cong 0$ and $\epsilon_{PP} \cong 0$. In addition, we also checked that the accuracy of these values increases with the number of sequences used for the training set and with the number of chosen decoys. Indeed, increasing the size of the training set (and the number of decoys) increases the number of inequalities to be satisfied (Eq. (5.14)). It follows that the vector $\vec{\epsilon}$ has to be more and more fine-tuned in order to satisfy all of them and consequently it converges toward the original interaction values.

In order to obtain a further proof of the reliability of the method, we furthermore apply our algorithm to a modified HP model. The idea is to check whether the perceptron is able to find the correct set of interaction values also in the case of non-degenerate values (in the standard HP model $\epsilon_{HP} = \epsilon_{PP} = 0$). In addition, we use non-commensurable interaction values so as avoiding any possible degeneracy. For two sets of different interaction values, we therefore design proteins and keep, as usual, those with a unique native state. Afterward the perceptron is applied to the two different sets.

We choose the first set of interaction values so that $\epsilon_{HH} \cong 2\epsilon_{HP}$:

$$\begin{aligned} \epsilon_{PP} &= 0 \\ \epsilon_{HP} = \epsilon_{PH} &= -0.446 \\ \epsilon_{HH} &= -0.895 \end{aligned} \tag{5.17}$$

and we fix the second set so as possessing a large difference between $\epsilon_{HH}$ and $\epsilon_{HP}$:

$$\begin{aligned} \epsilon_{PP} &= 0 \\ \epsilon_{HP} = \epsilon_{PH} &= -0.199 \\ \epsilon_{HH} &= -0.980 \end{aligned} \tag{5.18}$$

Protein design with these two sets of values gives rise to a set of 4 585 proteins for the first case, respectively 4 760 proteins for the second case. This difference is due to the fact that with the first set of interaction values, three HP interactions give a larger energy contribution as a single HH interaction. This is not the case for the second set of interaction values. It follows that the native states of the two sets are not completely overlapped.

|  | SET1 Original | SET 1 Perceptron | SET 2 Original | SET 2 Perceptron |
|---|---|---|---|---|
| $\epsilon_{PP}$ | 0 | -0.009 | 0 | 0 |
| $\epsilon_{HP}$ | -0.446 | -0.375 | -0.199 | -0.221 |
| $\epsilon_{HH}$ | -0.895 | -0.927 | -0.980 | -0.975 |

Table 5.1: Modified HP model: results for two different sets of interaction values. In both cases the perceptron is able to recover the original interaction values used for protein design: the rightness of the results does not depend on the chosen interaction values.

Tab. 5.1 shows the results obtained with the perceptron, using the whole set of sequences and 15 decoys for each sequence (chosen again in the first three excited states). Remarkably, in both cases, the perceptron recovers the original interaction values: the rightness of the results found with the HP model does not depend on a particular choice of interaction values.

These results indicate that a perceptron approach used to find effective interactions is reliable. Indeed, they demonstrate that if the *original* interactions that stabilize the native state are amino-acid interactions, the learning algorithm is able to recover them. In the next section, we apply the same method to the HPW model with the aim to express the solvent effects by means of effective amino-acid interactions.

### 5.2.4 The perceptron on the HPW model

To apply the method to proteins designed with the HPW model, we first have to define the effective interaction Hamiltonian that is supposed to replace the original one (the latter being only composed by solvent terms). However, since we do not know, *a priori*, what kind of contacts are able to replace the solvent effects, we are forced to deal with different kinds of Hamiltonians: we begin with the simplest description of energy and in case of failure of the method, *i.e.* if the perceptron is not able to find a set of interactions that stabilize the proteins, we increase the number of contacts. In our case, the simplest Hamiltonian is the HP one, since composed by only nearest-neighbors contacts:

$$H(S,\Gamma) = \sum_{i<j} \epsilon^{nn}_{p_i p_j} \Delta_{nn}(\vec{r}_i, \vec{r}_j), \tag{5.19}$$

with the contact matrix:

$$\Delta_{nn}(\vec{r}_i, \vec{r}_j) = \begin{cases} 1 & \text{if } i,j \text{ are } nn \\ 0 & \text{otherwise.} \end{cases} \tag{5.20}$$

where $nn = nearest\text{-}neighbors$. Same as for the HP model we deal with sequences of length $L = 16$: for each sequence in $\{S_i\}_{16}$, we take 15 decoys, randomly chosen from the first three excited states. As for the size of the training set, it is important to stress that in the case of HP proteins, we know *a priori* the values of the effective interactions, since they were used to design proteins. It follows that for any given training set, once the perceptron reaches a solution, we can compare the obtained values with the original ones. If there is agreement, the criterion of stability is therefore also automatically satisfied for all sequences not belonging to the training set. On the contrary, in the case of proteins of the HPW model, the presence of a solution for a given training set does not forcedly imply that these values are correct for any sequence. As a consequence, we begin to apply our perceptron algorithm to the *whole* set of sequences. The used algorithm is the same as the one we applied to proteins of the HP model. Once all vectors $\overrightarrow{C^\Delta}$ are computed using Eq. (5.13), starting from a random interaction vector $\vec{\epsilon}_0$, at each step $i$, the scalar product

$$\vec{\epsilon}_i \cdot \overrightarrow{C^\Delta}(S, \Gamma_d(S), \Gamma(S)) \qquad (5.21)$$

is computed for each sequence $S$. The vector $\overrightarrow{C_L^\Delta}$ with the lowest scalar product is identified and the new weight vector $\vec{\epsilon}_{i+1}$ becomes:

$$\vec{\epsilon}_{i+1} = \vec{\epsilon}_i + \eta \overrightarrow{C_L^\Delta} \qquad (5.22)$$

This process is iterated for a predefined number of steps $N$ and afterward the criterion of *learnability* is tested, reminding that if a solution exists, we have that:

$$\vec{\epsilon}_i \cdot \overrightarrow{C_L^\Delta}(S, \Gamma_d(S), \Gamma(S)) > 0 \qquad (5.23)$$

The result of this first approach is rather surprising: the problem is *unlearnable*. It is impossible to find a vector $\vec{\epsilon}_c$ satisfying *all* inequalities of Eq. (5.14): there is no interaction values able to stabilize all proteins at once. This result reveals that the solvent effects can not be described by a simple pairwise Hamiltonian.

While the determination of *global* interaction values was not successful, we furthermore investigate the learnability of *specific* interaction values, *i.e.* those that are at least able to stabilize small sets of sequences $\{S\}^i \subset \{S\}$. For this purpose, we use a new approach to the problem: instead of using the whole set of sequences, we limit our investigation on small sets. First, we apply the learning algorithm to two randomly chosen sequences. Then, at every new step, if a vector $\vec{\epsilon}_c$ exists satisfying all inequalities of Eq. (5.14), we add a new sequence and we re-apply the algorithm. For each run $i$, if the algorithm does not find a solution in a predefined number of steps $N$, we retain the size of the group $\{S\}^i$ and the related vector $\vec{\epsilon}_c^i$.

Fig. 5.6 shows the distribution of the sizes of these groups, computed on 1 000 runs. For each run $i$, the size of the group $\{S\}^i$ is represented on the $x$-axis in [%] of the whole set of sequences $\{S\}_{16}$ (the size of the latter being $M_{16} = 2\ 431$) and the distribution of these values is shown on the $y$-axis. From this picture it is clear that the number of parameters in an effective interaction Hamiltonian is too small if we use only nearest-neighbors contacts. Indeed, even in the best case, only $\sim 2\%$ of the total number of sequences can be stabilized by means of a common set of interaction values. This lack of success can be explained by the fact that with the Hamiltonian of Eq. (5.8) many sequences have

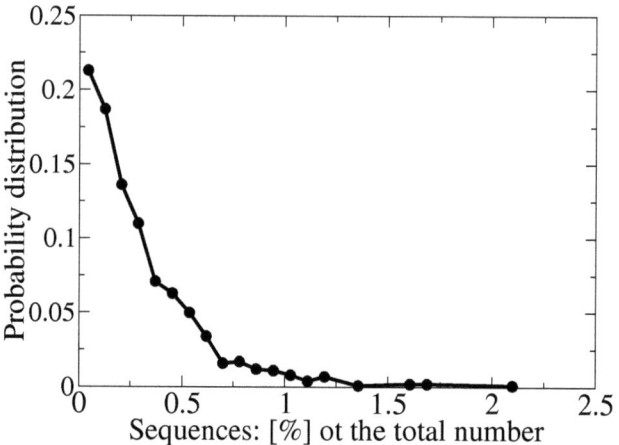

Figure 5.6: Probability distribution of the sizes of learnable groups computed with the *nearest-neighbors* Hamiltonian. For each run, the perceptron finds a set of values satisfying a maximal number of $x$ sequences, represented in [%] of the total on the $x$-axis. On the $y$-axis the probability distribution, computed on 1000 runs, is shown. Even in the best case, the percent of learnable sequences is only $\sim 2$.

at least one excited conformation possessing exactly the same number of contacts as its native state. It follows that the latter can not be recognized as native among the other conformations.

In order to increase the size of the set of learnable sequences we therefore introduce a new Hamiltonian, which takes into account also the *next-nearest neighbors* contacts, *i.e.* those between amino-acids located at the end of a diagonal of the square lattice:

$$H(S,\Gamma) = \sum_{i<j} \epsilon^{nn}_{p_i p_j} \Delta_{nn}(\vec{r}_i, \vec{r}_j) + \sum_{i<j} \epsilon^{nnn}_{p_i p_j} \Delta_{nnn}(\vec{r}_i, \vec{r}_j) + \quad (5.24)$$

with the contact matrices:

$$\Delta_{nn}(\vec{r}_i, \vec{r}_j) = \begin{cases} 1 & \text{if } i,j \text{ are } nn \\ 0 & \text{otherwise.} \end{cases}$$
$$\Delta_{nnn}(\vec{r}_i, \vec{r}_j) = \begin{cases} 1 & \text{if } i,j \text{ are } nnn \\ 0 & \text{otherwise.} \end{cases} \quad (5.25)$$

where $nnn = $ *next-nearest neighbors*.

For each sequence $S$, the vector $\overrightarrow{C^\Delta(S)}$ is recomputed using the new contact matrix of Eq. (5.25) and we apply directly the algorithm used for the search of *specific* interaction

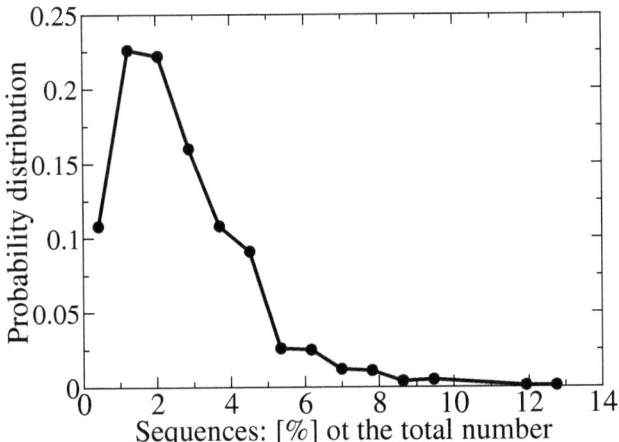

Figure 5.7: Probability distribution of the sizes of learnable groups computed with the *next-nearest neighbors* Hamiltonian. The addition of the next-nearest neighbors interactions increases the average number of learnable sequences. In some cases, the perceptron is able to find a set of values able to stabilize at once almost 13% of the total number of proteins.

values. Indeed, if a set of values exists able to stabilize the whole set of proteins, this algorithm stops only once this set is found, *i.e.* once all sequences are taken into account.

Fig. 5.7 shows the results obtained on 1 000 runs. As expected, the number of learnable sequences increases if next-nearest neighbors contacts are taken into account. Indeed, the curve is now peaked around 1.5% and the maximal number of sequences that can be stabilized with a unique set of interaction values is almost 13% of the total number of proteins. This increase of the size of learnable sequences is mainly obtained by the decrease of degeneracy of the number of contacts between the native state and the decoys. Decrease induced by the introduction of the next-nearest neighbors contacts. However, even these results show that the number of learnable sequences increases drastically by including the next-nearest neighbors contacts, it is still not possible to find a *global* set of interactions.

Since pairwise interactions are not able to reproduce correctly the solvent effects, we therefore introduce a further Hamiltonian (henceforth called *many-body* Hamiltonian),

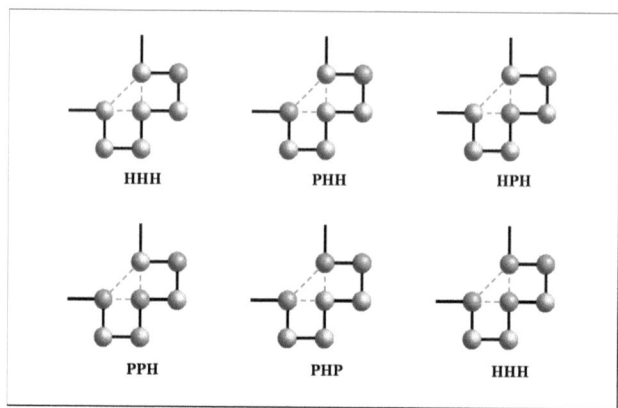

Figure 5.8: The six possible three-body interactions. Since the latter are composed by two nearest-neighbors contacts and one next-nearest neighbors contact, all possible permutations have to be taken into account. The two different colors represent the two kinds of amino-acids H and P, while the interactions are represented by red dashed lines.

which takes into account also the *three-body* interactions:

$$\begin{aligned}
H(S,\Gamma) &= \sum_{i<j} \epsilon^{nn}_{p_i p_j} \Delta_{nn}(\vec{r}_i, \vec{r}_j) + \\
&\quad \sum_{i<j} \epsilon^{nnn}_{p_i p_j} \Delta_{nnn}(\vec{r}_i, \vec{r}_j) + \\
&\quad \sum_{i<j<k} \epsilon^{3B}_{p_i p_j p_k} \Delta_{3B}(\vec{r}_i, \vec{r}_j, \vec{r}_k),
\end{aligned} \qquad (5.26)$$

with the contact matrices:

$$\begin{aligned}
\Delta_{nn}(\vec{r}_i, \vec{r}_j) &= \begin{cases} 1 & \text{if } i,j \text{ are } nn \\ 0 & \text{otherwise.} \end{cases} \\
\Delta_{nnn}(\vec{r}_i, \vec{r}_j) &= \begin{cases} 1 & \text{if } i,j \text{ are } nnn \\ 0 & \text{otherwise.} \end{cases} \\
\Delta_{3B}(\vec{r}_i, \vec{r}_j, \vec{r}_k) &= \begin{cases} 1 & \text{if } i,j,k \text{ are} \\ & 2 \ nn \text{ and } 1 \ nnn \\ 0 & \text{otherwise.} \end{cases}
\end{aligned} \qquad (5.27)$$

This Hamiltonian introduces six more dimensions into the vector space, the perceptron has therefore to deal with 12-dimensional vectors. Indeed, since three-body interactions are composed by two nearest-neighbors contacts and one next-nearest neighbors contact, we need to distinguish the groups with same amino-acid composition but different residue arrangements. The six possible permutations are pictorially shown in Fig. 5.8.

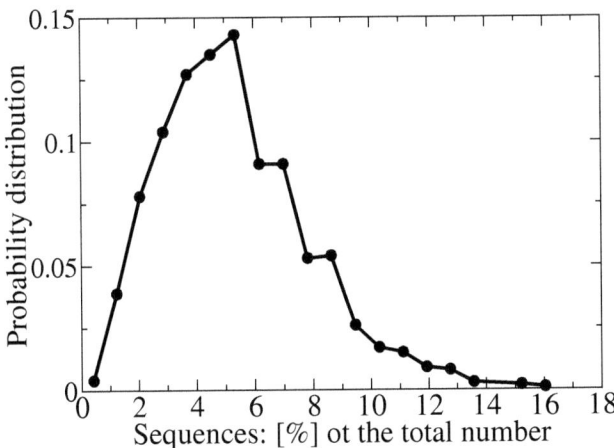

Figure 5.9: Probability distribution of the sizes of learnable groups computed with the *many-body* Hamiltonian on 1 000 runs. The addition of three-body interactions increases the average number of learnable sequences with respect to the pairwise Hamiltonians.

Same as for the other two Hamiltonians, we compute the vectors $\overrightarrow{C^\Delta}$ with the new contact matrix of Eq. (5.27) for all sequences (native state and 15 decoys) and we re-apply the specific interaction algorithm.

Fig. 5.9 shows the obtained values with the many-body Hamiltonian for 1 000 runs. As expected, the number of learnable sequences further increases by adding new interactions. The curve is peaked around 6%, while for the best case the perceptron finds a set of $\sim 16\%$ of sequences that can be stabilized with a unique set of interaction values. However, even if the size of learnable groups of proteins becomes important, it is still much smaller than the total number of sequences $M_{16}$.

In summary, from these results we can conclude that the high energy degeneracy implicit in the nearest-neighbors Hamiltonian is partially avoided introducing next-nearest neighbors contacts, and three-body interactions allow a more accurate tuning between the different potentials: the result becomes more robust and the typical size of a learnable set increases. However, it is not possible to find a set of global interaction values able to reproduce correctly the solvent effects of our model.

We are furthermore interested in the specific values of the different interactions. For this purpose, for each run $i$, we keep the last vector $\vec{\epsilon}_i$ for which the problem still have a solution. In order to investigate the relative importance of the different interactions, this is done for the many-body Hamiltonian. Fig. 5.10 shows the nearest-neighbors contact values, computed on 1 000 runs. Interestingly, the peaks of the curves satisfy the inequalities $\epsilon^{nn}_{HH} < \epsilon^{nn}_{HP} < \epsilon^{nn}_{PP}$. Indeed this result is expected from the energy defini-

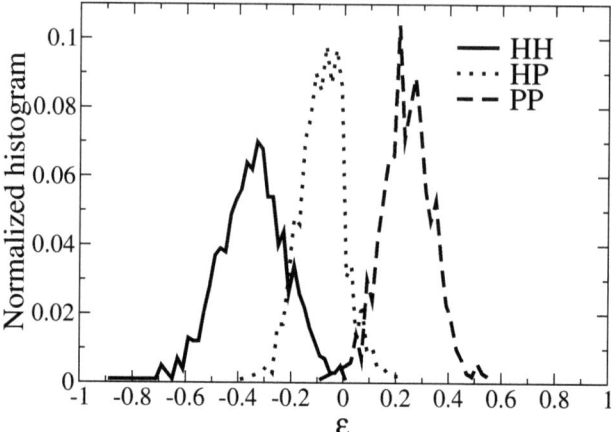

Figure 5.10: Distribution of the nearest-neighbors interaction values obtained on 1 000 runs. For each run $i$, once the algorithm stops because the problem becomes unlearnable, we keep the last interaction vector $\vec{\epsilon}$ for which the problem was still learnable. The peaks of the curves satisfy the inequalities $\epsilon_{HH}^{nn} < \epsilon_{HP}^{nn} < \epsilon_{PP}^{nn}$, as expected from the energy definition of the solvent.

tion of the solvent, since the energy gain of a hydrophobic-hydrophobic residue couple is higher than the one of a hydrophobic-polar couple (no energy gain occurs by sticking together two polar residues). The same behavior is also observed in the case of next-nearest neighbors contacts, since, as shown in Fig. 5.11, we have that $\epsilon_{HH}^{nnn} < \epsilon_{HP}^{nnn} < \epsilon_{PP}^{nnn}$. Moreover, this condition seems to be more essential for the next-nearest neighbors values as for the simple nearest-neighbors ones, since the distribution of the latter is broader.

Finally, Fig. 5.12 shows the value of the three-body interactions. Qualitatively all these interactions behave similarly: they are peaked around zero but exhibit a broad distribution. From these graphs, we can conclude that, in most of the cases, the perceptron tries to force to zero the three-body interactions, meaning that proteins are mainly stabilized by pairwise interactions. On the other hand, in some cases, particular three-body interaction values allow a fine-tuning of the interaction sets which rises to an increase of the number of learnable sequences by addition of the three-body interactions with respect to the pairwise Hamiltonians.

The aim of the former computations was to determine the average number of learnable proteins. It is however important to stress that the algorithm stops because it has not found a solution in $N$ steps, not because the problem has proven to be unlearnable. It follows that, in principle, the sizes of the sets may be larger as what shown with our computations. Indeed, we did not check for the absolute maximal number of learnable proteins, but, on the contrary, we were mainly interested in the average size of the groups and in the behavior of the interaction values. However, in order to proof that the problem

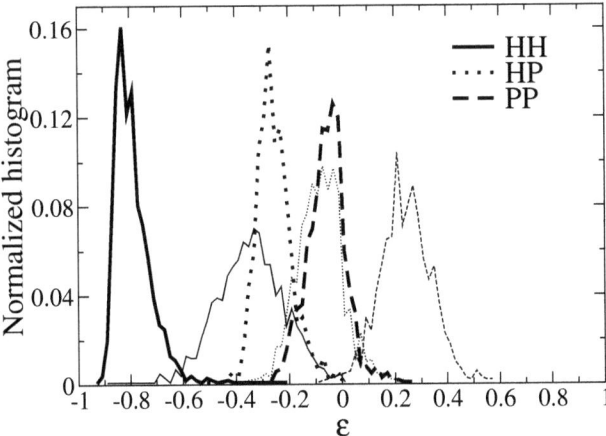

Figure 5.11: Distribution of the next-nearest neighbors interactions. Similar to the nearest-neighbors case, also these interactions satisfy the inequalities $\epsilon_{HH}^{nnn} < \epsilon_{HP}^{nnn} < \epsilon_{PP}^{nnn}$. For comparison, the nearest-neighbors values are replotted with thinner lines.

of finding a *global* set of interactions is indeed unlearnable, we also apply a modified algorithm, which was first introduced in [76] and applied to the crambin problem [17]. This algorithm proves unlearnability monitoring a *despair* parameter $d$: if this parameter $d$ exceeds a critical value $d_c$ after a *finite* number of steps, the problem is unlearnable. First, we apply this algorithm using the *whole* set of sequences and the many-body Hamiltonian, since the latter is the one that gives rise to the largest number of learnable sequences. As expected, the result is in agreement with what found in our former computations: the problem is *rigorously* unlearnable. In addition, we also check the groups of proteins generated by our original algorithm. It turns out that, when looking at the small sets $\{S\}^i$, we find that roughly 20% of them is indeed unlearnable, whereas the remaining 80% corresponds to the worst case of convergence for the theorem [76], case for which unlearnability, if any, could be proven only in $N \approx 10^{30}$ steps. Obviously, this number of steps is completely beyond our computational capacities. Yet, since we know that the whole set is unlearnable, we look at upper bounds for the sizes of the sets $\{S\}^i$: to each of them we add sequences till unlearnability is rigorous. We find that just a few extra sequences are enough and the peak of the histogram in Fig. 5.9 moves to about 8%.

Even if effective interaction Hamiltonians with an appropriate set of interaction values are able to recognize the native state of protein, no information is given about the energy landscape. However, since effective Hamiltonians may also be employed to investigate dynamics of proteins, the criterion of similarity between the energy landscape of the original Hamiltonian and the one of the effective Hamiltonian should also be respect. Indeed, since the common picture of the energy landscape of proteins is that the native state lies at the bottom of a funnel, and folding proceeds along its walls, scrambling the

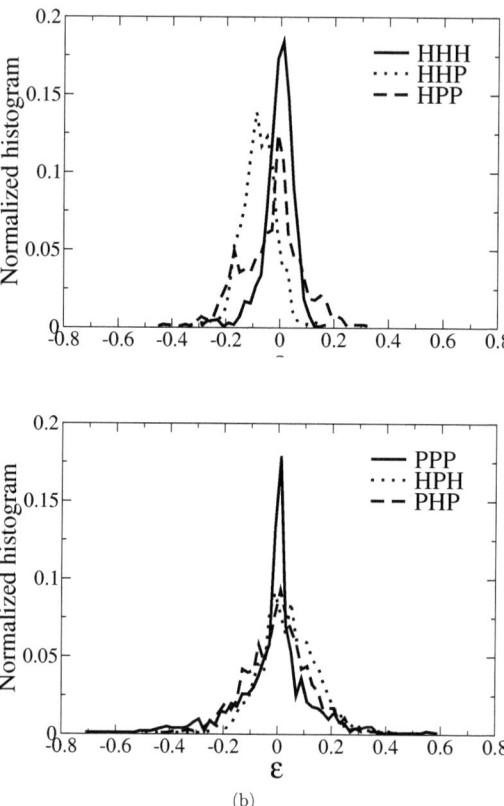

(b)

Figure 5.12: Distribution of the three-body interaction values. The perceptron tries to force these values to zero. However, in some cases, a non-zero value of one or more of these interactions gives rise to a better determination of the native state, since the size of learnable sequences increases by adding three-body interactions to the next-nearest neighbors Hamiltonian. The legend of the different interaction values is referred to the one of Fig. 5.8: (a) HHH, HHP, HPP. (b) PPP, HPH, PHP.

energies of the excited states can affect dramatically the dynamics. In order to obtain some information about this further crucial point, we therefore investigate the excited states given by the effective Hamiltonian of Eq. (5.26) in the many-body case. For this purpose, we recompute, for different sequences $S_i$, the energies of the conformations in the three first excited states (according to the explicit solvent Hamiltonian of the HPW mode.

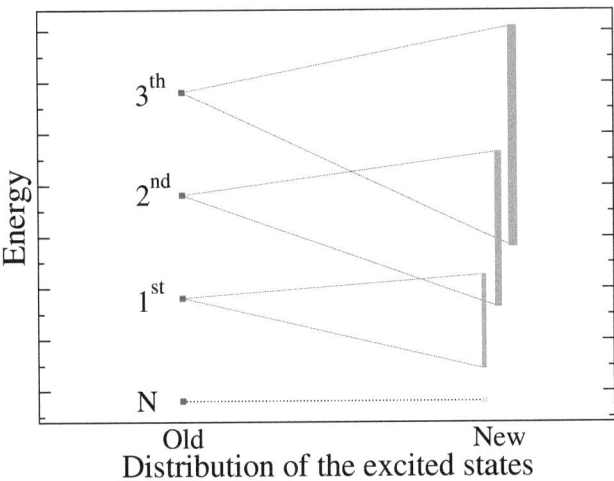

Distribution of the excited states

Figure 5.13: Distribution of the excited states. The energy levels for the solvent dependent Hamiltonian are represented on the *l.h.s.* of the picture. For every configuration the new energy is computed, using the interaction potentials found by the perceptron. The typical behavior for a given sequence is shown on the *r.h.s.*. The introduction of the potentials destroys the initial energy hierarchy: the states are overlapped and the energy landscape changes.

Fig. 5.13 shows the typical behavior obtained by these computations. It turns out that for all sequences the initial hierarchy (left side in Fig. 5.13) of the excited states imposed by the solvent dependent Hamiltonian changes if we use an effective interaction description of the energy (right side in Fig. 5.13). The picture shows clearly that the use of effective interactions destroys the initial energy hierarchy. The originally separated excited states of the HPW Hamiltonian become overlapped if expressed with an effective interaction Hamiltonian. As a consequence, we can select two different conformations $\Gamma_i$ and $\Gamma_j$, so that the inequality relating their two energies is $E^{HPW}(\Gamma_i) < E^{HPW}(\Gamma_j)$ using the solvent dependent Hamiltonian, but, on the contrary, it is reversed using the effective one of Eq. (5.26), so that $E^{eff}(\Gamma_i) > E^{eff}(\Gamma_j)$. Moreover, even if the native state is a global minimum of the energy also for the effective interaction Hamiltonian, this energy scrambling may result in a change of the intermediate states, as well as of the folding pathway (pictorially shown in Fig. 5.15). This result suggest that, even for restricted protein sets, some care should be taken when using effective potentials for dynamical studies.

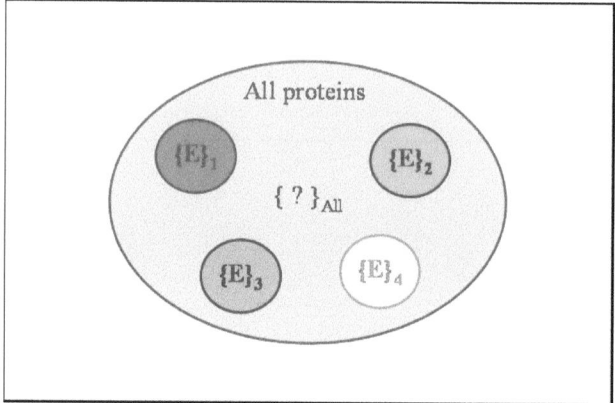

Figure 5.14: Schematic representation of the results obtained with the perceptron applied on HPW proteins. Different groups of proteins (represented by different colors) give rise to different sets of interaction values $\{E\}_i$. The latter are only able to stabilize proteins belonging to their own sets. No set of *global* effective interactions exists able to stabilize all proteins at once.

In summary, in this chapter we used a common method to extract effective interactions on real proteins: the perceptron method. We first applied this algorithm to HP proteins in order to check the validity of the method. It turned out that, if the *original* interactions used to design protein are amino-acid interactions, the perceptron is able to recover them. On the other hand, the method applied to HPW proteins revealed that the solvent effects can not be reproduced by effective residue interactions. Indeed, it turned out that proteins are divided into groups: for each group we found interactions values able to recognize the correct native state of the sequences inside this group (pictorially shown in Fig. 5.14). Yet it is not possible to determine *global* values, meaning that the solvent effects are more complex than simple effective amino-acid interactions. Moreover, the energy landscapes determined by the new effective Hamiltonians are different from the ones of the solvent dependent Hamiltonian. As a consequence, the dynamical folding process may change too. We therefore suggest that these two main results may explain why it is not possible to determine *global* effective interactions for real proteins.

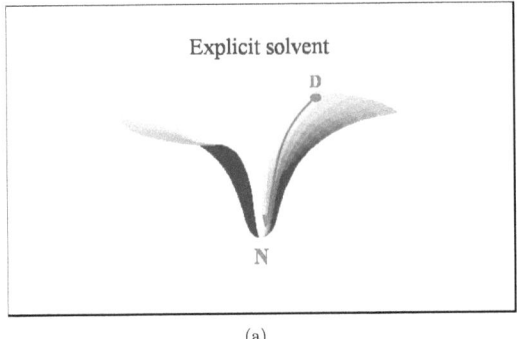

(a)

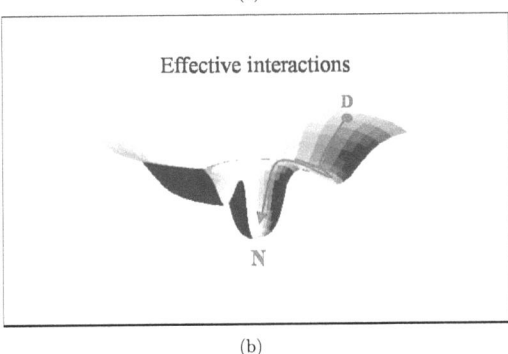

(b)

Figure 5.15: Schematic representation of the change in the energy landscape. The scrambling in the energy levels (represented in Fig. 5.13) induced by the effective interaction Hamiltonian may result in a change of the folding pathway. (a) The original funnel obtained with the solvent dependent Hamiltonian. (b) A possible energy landscape generated by the new effective interaction Hamiltonian.

# Chapter 6
# Summary

This book has shown that many qualitative features of proteins can be reproduced by a simple lattice model with a (semi)explicit description of the solvent. The thoroughly investigation of proteins designed with the HPW model evidenced that the hydrophobic effect induced by the solvent is crucial for the protein behavior. Indeed, the HPW model, focusing on the solvent effects, is able to reproduce many properties of real proteins. For instance, it turned out that HPW proteins possess a very compact native state characterized by the presence of a hydrophobic core. These structures are typical of globular proteins. Moreover, the model reproduces correctly both *cold* and *warm* denaturation, behavior that is not reproduced by most Hamiltonians used in protein folding, since their native state is the $T = 0$ ground state (GS) of the model.

As for statistical quantities, it has been found that not all structures possess the same degree of designability but, on the contrary, sequences have a preference for some particular conformations (*i.e.* the *high designable* structures). This result is in agreement with the concept of protein *folds* used as a taxonomical classification of real proteins.

Some insights on the influence of chaotropic agents on protein stability have also been given. By modifing the MLG model so as reproducing the presence of chaotropes, it has been shown that the alteration of the hydrogen-bond networks of water molecules due to the presence of chaotropic agents induces destabilization of proteins, and leads, in the extreme case, to complete denaturation. These effects are also observed in experiments and commonly used to denaturate proteins. In addition, one found that the presence of chaotropes originates an effective interaction between the chaotropes and the protein. As a result, the chaotropic molecules accumulate near the surface of the protein. This effect is due to the disruption of the hydrogen-bond networks of the water and no direct cosolvent-protein interaction is needed to reproduce it (even if this interaction is likely to occur in the case of real proteins).

The validity of effective amino-acid interaction Hamiltonians for protein folding and protein stability is still an open issue. Indeed, all attempts to find a set of *global* interaction values able to stabilize all proteins at once failed. Using the same technique used for real proteins, namely a perceptron algorithm, one therefore tested the validity of this approach. The aim of this investigation was to generate an effective interaction Hamiltonian able to recognize the native states of the HPW proteins. However, it turned out that the solvent effects can not be reproduced by effective residue interactions. Indeed, the results have shown that proteins are divided into groups, each of them possessing their own *specific* set of interaction values. Yet it is not possible to determine *global* values able to stabilize all proteins at once. These results demonstrate that the solvent effects are more complex than simple effective amino-acid interactions. Moreover, it has also been found that the energy landscapes determined by the new effective Hamiltonians are different from the ones of the solvent dependent Hamiltonian. This difference may therefore lead to a modification of the folding pathway and consequently of the dynamical folding process. We therefore argue that, even if specific effective interactions are efficient to stabilize specific proteins, some care should be taken when using effective potentials for dynamical studies, since the energy landscape may not be correctly reproduced.

In summary, of particular relevance is our observation that a simple model, in which the potential energy is described in terms of the interactions between amino-acids only, does not correctly reproduce solvent effects. Approaches like this, in which solvent effects are treated implicitly, are commonly employed in many of more sophisticated models for protein folding dynamics. Our results are then of great importance as they suggest the treatment of the solvent in these models may need to be re-examined.

# Bibliography

[1] D. Voet and J. G. Voet, *Biochemistry - Second Edition* (John Wiley & Sons, Inc., New York, 1995).

[2] http://www.globalrph.com/

[3] http://www.artwiredmedia.com/

[4] http://www.rcsb.org/pdb/

[5] http://hekto.med.unc.edu:8080/XRAY/

[6] Bongini L, Fanelli D, Piazza F, *et al.*, Proc. Natl. Acad. Sci. USA **101**, 6466 (2004).

[7] C. B. Anfinsen, Science **181**, 223 (1973).

[8] C. Levinthal, J. Chim. Phys. **65**, 44 (1968).

[9] D.B. Wetlaufer, Proc. Natl. Acad. Sci. USA **70**, 691 (1973).

[10] K. A. Dill, Protein Science **8**, 1166 (1999).

[11] U. Mayor, N.R. Guydosh, C.M. Johnson *et al.*, Nature **421**, 863 (2003).

[12] S. Lifson and C. Sander, Nature **282**, 109 (1979).

[13] S. Lifson and C. Sander, J. Mol. Biol., **139**, 627 (1980).

[14] S. Miyazawa and R.L. Jernigan, Macromolecules **18**, 534 (1985).

[15] V.N. Maiorov and G.M. Crippen, J. Mol. Biol. **227**, 876 (1992).

[16] F. Rosenblatt, *Principles of Neurodynamics*, Spartan Books, New York, (1962).

[17] M. Vendruscolo, E. Domany, J. Chem Phys. **109**, 11101 (1998).

[18] M. Vendruscolo, R. Najmanovich and E. Domany, Proteins **38**, 134 (2000).

[19] W. Kauzmann, Adv. Protein Chem. **14**, 1 (1959).

[20] J. W. McBain, *Colloid Science*, (D.C. Heath, Boston, 1950).

[21] P. Debye, Ann. N.Y. Acad. Sci. **51**, 575 (1949).

[22] H. S. Frank and M. W. Evans, J. Chem. Phys. **13**, 507 (1945).

[23] P. M. Wiggins, Microbiol. Rev. **54**, 432 (1990).

[24] G. L. Pollack, Science **251**, 1323 (1991).

[25] R. Ludwig, Angew. Chem. Int. Ed. **40**, 1808 (2001).

[26] A. Ben-Naim, *Water and Aqueous Solutions*, (Plenum Press, New York and London, 1980).

[27] P. L. Privalov and S. J.. Gill, Adv. Protein Chem. **39**, 191 (1988).

[28] A. Perstemilidis, A. Saxena, A. K. Soper, T. Head-Gordon and R. M.Glaeser, Proc. Natl. Acad. Sci. USA **93**, 10769 (1996).

[29] Y. C. Bae, S. M. Lambert, D.S. Soane and J. M. Prausnitz, Macromolecules **24**, 4403 (1991).

[30] H. G. Schild and D. A. Tirrel, J. Phys. Chem. **94**, 4352 (1990).

[31] S. H. Yalkowsky, *Solubility and Solubilization in Aqueous Media*, (American Chemical Society and Oxford University Press, New York, 1999).

[32] P. L. Privalov, CRC Crit. Rev. Biochem. Mol. Biol. **25**, 181 (1990).

[33] K. P. Murphy, P. L. Privalov and S. J. Gill, Science **25**, 559 (1990).

[34] G. I. Makhatatadze and P. L. Privalov, J. Mol. Biol. **213**, 375 (1990).

[35] G. I. Makhatatadze and P. L. Privalov, J. Mol. Biol. **232**, 639 (1993).

[36] P. L. Privalov and G. I. Makhatatadze, J. Mol. Biol. **232**, 660 (1993).

[37] G. I. Makhatatadze and P.L. Privalov, Adv. Protein Chem. **47**, 307 (1995).

[38] K. P. Murphy, P. L. Privalov and S. J. Gill, Science **25**, 559 (1990).

[39] N. C. Pace and C. Tanford, Biochemistry **7**, 198 (1968).

[40] P. L. Privalov, Yu. V. Griko, S. Yu. Venyaminov and V. P.Kutyshenko, J. Mol. Biol. **190**, 487 (1986).

[41] A. R. Fersht, *Structure and Mechanism in Protein Science: A Guide to Enzyme Catalysis and Protein Folding* (W. H. Freeman & Co., New York, 1999).

[42] H. S. Frank and F. Franks, J. Chem. Phys. **48**, 4746 (1968).

[43] D. O. V. Alonso and K. A. Dill, Biochemistry **30**, 5974 (1991).

[44] A. Caflisch and M. Karplus, Proc. Natl. Acad. Sci. USA **91**, 1746 (1994).

[45] J. A. Schellman, Biopolymers **14**, 999 (1994).

[46] A. Wallqvist, D. G. Covell, and D. Thirumalai, J. Am. Chem. Soc. **120**, 427 (1998).

[47] R. Chitra and P. E. Smith, J. Phys. Chem. B **105**, 11513 (2001).

[48] K. Ting and R. Jernigan, Proc. Natl. Acad. Sci. USA **99**, 9721 (2002).

[49] J. A. Schellman, Biophys. Chem. **96**, 91 (2002).

[50] S. Shimizu and H. S. Chan, Proteins **49**, 560 (2002).

[51] D. Tobi, R. Elber, and D. Thirumalai, Biopolymers **68**, 359 (2003).

[52] B. J. Bennion and V. Daggett, Proc. Natl. Acad. Sci. USA **100**, 5142 (2003).

[53] P. De Los Rios and G. Caldarelli, Phys. Rev. E **62**, 8449 (2000).

[54] G. Caldarelli and P. De Los Rios, J. Biol. Phys. **27**, 229 (2001).

[55] K. F. Lau and K. A. Dill, Macromolecules **22**, 3986 (1989).

[56] H. S. Chan and K. A. Dill, Phys. Today **46**, 24 (1993).

[57] N. Muller, Acc. Chem. Res. **23**, 23 (1990).

[58] B. Lee and G. Graziano, J. Am. Chem. Soc. **22**, 5163 (1996).

[59] C. Tanford, Science **200**, 1012 (1978);

[60] K.A.T. Silverstein, A.D.J. Haymet, and K.A. Dill, J. Chem. Phys. **111**, 8000 (1999).

[61] F. Y. Wu, Rev. Mod. Phys. **54**, 235 (1982).

[62] http://www.expasy.org/sprot/

[63] D. Eisenberg *et. al.*, J. Mol. Biol. **179**, 125 (1984).

[64] J.L. Cornette *et al.*, J.Mol.Biol. **196**, 659 (1987).

[65] http://www2.ebi.ac.uk/dali/fssp/

[66] S.H. White and R.E. Jacobs, J. Mol. Evol. **36**, 79 (1993).

[67] A somewhat related interpretation was obtained by direct inspection of HP proteins by C. T. Shih *et al.*, Phys. Rev. Lett. **84**, 386 (2000).

[68] U. K. Spiro, Z. Physiol. Chem. **30**, 182 (1900).

[69] J. P. Greenstein, J. Biol. Chem. **125**, 501 (1938).

[70] A. Torcini, R. Livi and A. Politi, cond-mat/0103270.

[71] K. A. Dill and S. Bromberg, *Molecular driving forces* (Garland Science, New York and London, 2003).

[72] C. N. Pace and K. W. Shaw, Proteins **4**, 1 (2000).

[73] J. K. Myers, C. N. Pace, and J. M. Scholtz, Prot. Sci. **4**, 2138 (1995).

[74] A. Caflisch and M. Karplus, Structure **7**, 477 (1999).

[75] W. Krauth and M. Mezard, J. Phys. A **20**, L745 (1987).

[76] D. Nabutovsky and E. Domany, Neural Computation **3**, 604 (1991).

# Scientific Publishing House
### offers
# free of charge publication

of current academic research papers, Bachelor´s Theses, Master's Theses, Dissertations or Scientific Monographs

If you have written a thesis which satisfies high content as well as formal demands, and you are interested in a remunerated publication of your work, please send an e-mail with some initial information about yourself and your work to *info@vdm-publishing-house.com*.

**Our editorial office will get in touch with you shortly.**

**VDM Publishing House Ltd.**
Meldrum Court 17.
Beau Bassin
Mauritius
www.vdm-publishing-house.com

Printed by Books on Demand GmbH, Norderstedt / Germany